FOREIGN OIL DEPENDENCE

FOREIGN OIL DEPENDENCE

Other books in the At Issue series:

FOREIGN OIL DEPENDENCE

James Haley, *Book Editor*

Bonnie Szumski, *Publisher*
Scott Barbour, *Managing Editor*
Helen Cothran, *Senior Editor*

GREENHAVEN
PRESS®

San Diego • Detroit • New York • San Francisco • Cleveland
New Haven, Conn. • Waterville, Maine • London • Munich

LIBRARY OF CONGRESS CATALOGING-IN-PUBLICATION DATA

Foreign oil dependence / James Haley, book editor.
    p. cm. — (At issue)
Includes bibliographical references and index.
ISBN 0-7377-2272-X (alk. paper) — ISBN 0-7377-2273-8 (pbk. : alk. paper)
    1. Petroleum industry and trade—Government policy—United States.
2. Petroleum industry and trade—Political aspects—United States. 3. Petroleum conservation—United States. I. Haley, James, 1968– . II. At issue (San Diego, Calif.)
HD9566.F67 2004
333.8'232'0973   2003066264

Printed in the United States of America

Contents

Introduction

The United States is the world's largest importer and user of oil. Daily, Americans consume nearly 21 million barrels of oil to power automobiles, generate electricity, and heat and cool homes. According to Henry S. Rowen, a researcher with the Hoover Institution, a conservative think tank, "With the same size economy, the European Union uses half as much motor fuel as we do." Given such a ravenous appetite for oil, it is not surprising that nearly 60 percent of the oil needed to keep the U.S. economy running is imported from foreign suppliers. More than 20 percent of that imported oil comes from the Persian Gulf region of the Middle East, which includes Iraq, Iran, Kuwait, Saudi Arabia, and the United Arab Emirates. Saudi Arabia alone sits on nearly one-fourth of the world's oil, with more than 261 billion barrels of proven reserves. Iraq has the world's second largest supply of oil, with an estimated 112 billion barrels.

America's heavy reliance on oil imported from the Persian Gulf has presented significant foreign policy challenges. The September 11, 2001, terrorist attacks have strained relations with Saudi Arabia, a nation ruled by an Islamic monarchy and the source of nearly 15 percent of the oil imported by the United States. Fifteen of the nineteen hijackers involved in the September 11 attacks were discovered to have been Saudi citizens. The U.S. government finds itself walking a diplomatic tightrope: Sanctioning the Saudi government for allowing terrorists to flourish might jeopardize the flow of oil to the United States and weaken the U.S. economy. Meanwhile, the Saudi government has offered little support to President George W. Bush's war against terrorism. Says Gawdat Bahgat, the director of the Center for Middle Eastern Studies,

> Some members in the U.S. Congress and several American news organizations have publicly criticized the lack of Saudi support in the war against terror. Their criticism has focused on the involvement of Saudi citizens in the terrorist attacks and on allegations that Saudi private money had been funneled to terrorist organizations. . . . In short, the Saudi record of cooperation with the United States in the war against terrorism is mixed.

More recently, antiwar protesters throughout Europe and the United States questioned the motives behind the 2003 U.S.-Britain invasion of Iraq, conducted without the approval of the United Nations (UN). The preemptive strike against Iraq was ostensibly based on U.S. intelligence indicating that Iraqi president Saddam Hussein was developing a program of weapons of mass destruction—including nuclear and chemical weapons—that posed an imminent threat to the security of the United States. A secondary contention voiced by the Bush administration was that Hussein had funneled money to al-Qaeda, the terrorist group re-

9

sponsible for the September 11 attacks against the United States. When questioned by the media, U.S. secretary of defense Donald Rumsfeld adamantly denied that the Iraq war was about oil, asserting that "it has nothing to do with oil, literally nothing to do with oil."

Notwithstanding Bush administration officials' protestations to the contrary, critics view the Iraq war as part of a long-term strategy by the U.S. government to dominate the oil-rich Persian Gulf region in defense of "oil security." This policy is based on the idea that unimpeded access to oil is of vital importance to U.S. security and economic health and should be secured by military means if necessary. Journalist Justin Fox explains that pursuing a policy of oil security means "shielding the U.S. from the economic damage caused by sudden, sharp rises in oil prices [as a result of] . . . wars, terrorism, accidents, natural disasters, or political decisions [that] might suddenly and significantly decrease the supply of oil and send prices skyrocketing."

While questions remain over what compelled President Bush to invade Iraq in 2003, a review of the historical record reveals that increasing U.S. dependence on imported oil has been followed by a corresponding willingness to interfere in the political affairs of the Persian Gulf countries. Many observers would agree with the assertion of Robert E. Ebel, the director of the energy program at the Center for Strategic and International Studies (CSIS), that oil is not treated by the U.S. government as simply another commodity. Contends Ebel, "Oil is high-profile stuff. Oil fuels military power, national treasuries, and international politics. It is no longer a commodity to be bought and sold within the confines of traditional energy supply and demand balances. Rather, it has been transformed into a determinant of well-being, of national security, and of international power."

Oil's transformation from a commodity to a determinant of national security and power began shortly after enormous oil fields were discovered in the Persian Gulf region in the late 1940s. In the years following World War II, the U.S. economy had become heavily reliant on cheap sources of oil to maintain an increasingly suburban lifestyle. Americans began to live farther from city centers, necessitating longer commutes to work, and moved to hot, arid parts of the country made habitable by energy-consuming air conditioners. Although domestic production of oil in places like Texas and Oklahoma supplied a significant portion of the oil consumed by Americans, geologists in the early 1950s had predicted—accurately, it turns out—that domestic oil reserves would peak around 1970.

Convinced that the security of the U.S. economy would someday hinge on a steady and cheap supply of foreign oil, President Harry Truman, who served as president from 1945 to 1953, stepped up U.S. political influence in the Persian Gulf through the use of covert operations conducted by the Central Intelligence Agency (CIA). In 1953 the CIA instigated the overthrow of Iranian premier Mohammed Mossadegh, whose nationalization of Iran's large oil fields had antagonized the United States. In his place arrived the American-friendly Shah, Mohammed Reza Pahlavi, who made America feel secure in its access to Iranian oil. The Shah reigned until 1979, when he found himself the victim of simmering resentment against the United States; he was overthrown by an Islamic fundamentalist revolution that installed the Ayatollah Khomeini as Iran's new leader.

Khomeini renationalized Iran's oil industry and fifty-two American embassy employees were held hostage by Iranian student radicals for more than a year. The U.S. government has considered Iran a dangerous "rogue" state ever since.

By the early 1970s, the United States faced other foreign policy challenges in the region related to U.S. support for the Middle Eastern nation of Israel. From October 1973 to January 1974, a boycott on oil sales to the United States was imposed by the Organization of Petroleum Exporting Countries (OPEC) in retaliation against U.S. support for Israel's October 6, 1973, invasion of Egypt and Syria. OPEC—a cartel of oil-producing nations whose members include Nigeria, Venezuela, Indonesia, and the Persian Gulf nations—had formed in September 1960 to exercise more control over world oil prices. As a result of the embargo, the United States experienced a severe "oil shock" during which oil prices skyrocketed from $5.12 a barrel to $11.65. Author Robert Gramlin contends in his book *Oil on the Edge* that "these increases were only the beginning of a steady climb that would see crude oil peak at over $30.00 a barrel seven years later."

OPEC's power in controlling world oil prices and its ability to wield oil as a political "weapon" was acutely felt: The lingering effects of the oil embargo wreaked havoc on the U.S. economy throughout the 1970s, leading to long gas lines, higher prices at the pump, runaway economic inflation, and high unemployment. The bleak domestic situation prompted President Jimmy Carter, who took office in 1977, to officially designate U.S. access to oil a matter of national security. As journalist Robert Dreyfuss explains,

> In January 1980, President Carter effectively declared the [Persian] Gulf a zone of U.S. influence, especially against encroachment from the Soviet Union. "Let our position be clear," he said, announcing what came to be known as the Carter Doctrine. "An attempt by any outside force to gain control of the Persian Gulf region will be regarded as an assault on the vital interests of the United States of America, and such an assault will be repelled by any means necessary, including military force." To back up this doctrine, Carter created the Rapid Deployment Force, an "over-the-horizon" military unit capable of rushing several thousand U.S. troops to the Gulf in a crisis.

President Ronald Reagan, Carter's successor, transformed the Rapid Deployment Force into the Central Command, "a new U.S. military command authority with responsibility for the [Persian] Gulf and the surrounding region from eastern Africa to Afghanistan," according to Dreyfuss.

During the 1990s, Middle East oil continued to influence U.S. foreign policy. On August 2, 1990, Iraq, under the leadership of President Saddam Hussein, invaded neighboring Kuwait. Iraq's unprovoked occupation prompted Saudi Arabia, which feared that an Iraqi invasion was imminent, to allow a permanent U.S. military presence on its soil. Then, in the spring of 1991, President George H.W. Bush ordered U.S. troops to expel Iraqi forces from Kuwait in a military campaign known as the Persian Gulf War. While the war was ostensibly about restoring freedom to the

Kuwaiti people, the threat of an Iraqi dictatorship controlling so much of the world's oil supply was clearly a factor behind Bush's decision to commit U.S. forces. The war marked the first time that U.S. military troops were used when the oil security of the United States was perceived to be at risk. Hussein was expelled from Kuwait, but he remained in power until President George W. Bush, the son of George H.W. Bush, commanded the invasion of Iraq in the spring of 2003.

For the sake of oil security, the United States has adopted an aggressive, and at times, militaristic foreign policy to ensure a steady supply of oil from the Persian Gulf. Some observers contend, however, that OPEC's purported ability to manipulate world oil prices—in effect, using oil as a political and economic weapon—has been exaggerated by the U.S. government. According to this view, oil should not be elevated to the status of a national security issue worthy of military intervention; after all, oil is just another commodity, like coffee, that is only valuable if it can be sold according to the laws of supply and demand. Explain Jerry Taylor and Peter Van Doren of the libertarian Cato Institute,

> The oil weapon is a myth and belief in that myth is crippling U.S. foreign policy. . . . OPEC is hardly in a position to punish the industrialized nations with a radical production cutback. . . . Because there's no other source of revenue for these economies other than oil, a major production cutback would bankrupt the OPEC countries and almost certainly trigger revolutions.

Another criticism of the U.S. government's current oil-security policy derives from the nature of the world oil market. Argues Taylor, "Even if every drop of oil we consumed came from Oklahoma, Texas, and Alaska, a cutback in OPEC production would raise domestic oil prices as high as if all our oil came from Saudi Arabia. That's because there are no regional markets for oil—only global markets—and regional prices invariably rise to the world price." According to Taylor, oil will fetch the same price no matter where it originates, so worrying about overreliance on foreign oil and expending capital and human resources to secure oil from unstable parts of the world is a wasted effort. Following this line of argument, Shibley Telhami, a senior fellow in the Brookings Institution Foreign Policy Studies program, contends that the cost of maintaining a military and political presence in the Persian Gulf is exceedingly expensive. Says Telhami, "Maintaining the U.S. military presence in the Persian Gulf costs upward of $60 billion a year. . . . One wonders why the United States devotes so much of its resources, energies, and war planning to the Persian Gulf. Would it not be more sensible to leave the oil issue to market forces and to leave politics out of it?"

Finally, as evidenced by the heated inquiry into what factors compelled the Bush administration to invade Iraq in 2003, critics of that war argue that pursuing a foreign policy closely aligned with protecting the nation's oil security undermines the credibility of the U.S. government abroad. Citizens and leaders of other countries, particularly in the European Union, have come to view U.S. policy in the Persian Gulf region as inconsistent with the U.S. government's supposed concern for discouraging undemocratic regimes around the world. For example, in the case of

Iraq, the Bush administration stated that freeing the Iraqi people from President Saddam Hussein's dictatorial rule was an important goal; meanwhile, the Saudi Arabian regime, which is thoroughly undemocratic but friendlier to American interests, receives scant criticism from the U.S. government. As U.S. dependence on foreign oil continues unabated, it may become ever more apparent to the rest of the world that oil is a crucial factor driving U.S. foreign policy. Says Lutz C. Kleveman, a reporter with the British journal the *Ecologist*, "As long as . . . the industrialized world's dependency on Middle Eastern oil continues unabated, conflicts are likely to break out which are essentially about securing the earth's remaining energy reserves." The following viewpoints in *At Issue: Foreign Oil Dependence* examine the controversial influence of oil on American foreign policy as well as debate ways in which the United States can reduce its dependence on foreign oil.

1

A World Oil Shortage Is Inevitable

Kenneth S. Deffeyes

Kenneth S. Deffeyes is a professor of geology at Princeton University in Princeton, New Jersey, and has worked as a field geologist for major oil companies.

The predictions of geologist M. King Hubbert, delivered in the 1950s, that global oil production will peak between 2004 and 2008 are proving to be accurate. The slowdown in oil production and the dwindling oil supply around the world will cause price increases and may precipitate a serious energy crisis. Major oil companies and oil-producing countries are not facing up to the problem. Planning for increased energy conservation and developing alternative energy sources should begin as soon as possible to avoid the economic impact of steeply rising oil prices.

G lobal oil production will probably reach a peak sometime during this decade [2001–2010]. After the peak, the world's production of crude oil will fall, never to rise again. The world will not run out of energy, but developing alternative energy sources on a large scale will take at least 10 years. The slowdown in oil production may already be beginning; the current price fluctuations for crude oil and natural gas may be the preamble to a major crisis.

The dwindling oil supply

In 1956, the geologist M. King Hubbert predicted that U.S. oil production would peak in the early 1970s. Almost everyone, inside and outside the oil industry, rejected Hubbert's analysis. The controversy raged until 1970, when the U.S. production of crude oil started to fall. Hubbert was right.

Around 1995, several analysts began applying Hubbert's method to world oil production, and most of them estimate that the peak year for world oil will be between 2004 and 2008. These analyses were reported in some of the most widely circulated sources; *Nature, Science,* and *Scientific*

American. None of our political leaders seem to be paying attention. If the predictions are correct, there will be enormous effects on the world economy. Even the poorest nations need fuel to run irrigation pumps. The industrialized nations will be bidding against one another for the dwindling oil supply. The good news is that we will put less carbon dioxide into the atmosphere. The bad news is that my pickup truck has a 25-gallon tank.

The experts are making their 2004–8 predictions by building on Hubbert's pioneering work. Hubbert made his 1956 prediction at a meeting of the American Petroleum Institute in San Antonio [Texas] where he predicted that U.S. oil production would peak in the early 1970s. He said later that the Shell Oil head office was on the phone right down to the last five minutes before the talk, asking Hubbert to withdraw his prediction. Hubbert had an exceedingly combative personality, and he went through with his announcement.

I went to work in 1958 at the Shell research lab in Houston [Texas] where Hubbert was the star of the show. He had extensive scientific accomplishments in addition to his oil prediction. His belligerence during technical arguments gave rise to a saying around the lab, "That Hubbert is a bastard, but at least he's our bastard." Luckily, I got off to a good start with Hubbert, he remained a good friend for the rest of his life.

Critics had many different reasons for rejecting Hubbert's oil prediction. Some were simply emotional; the oil business was highly profitable, and many people did not want to hear that the party would soon be over. A deeper reason was that many false prophets had appeared before. From 1900 onward, several of these people had divided the then known U.S. oil reserves by the annual rate of production. (Barrels of reserves divided by barrels per year gives an answer in years.) The typical answer was 10 years. Each of these forecasters started screaming that the U.S. petroleum industry would die in 10 years. They cried "wolf." During each ensuing 10 years, more oil reserves were added, and the industry actually grew instead of drying up. In 1956, many critics thought that Hubbert was yet another false prophet. Up through 1970, those who were following the story divided into pro-Hubbert and anti-Hubbert factions. One pro-Hubbert publication had the wonderful title "This Time the Wolf Really *Is* at the Door."

[Most experts] estimate that the peak year for world oil [production] will be between 2004 and 2008.

Hubbert's 1956 analysis tried out two different educated guesses for the amount of U.S. oil that would eventually be discovered and produced by conventional means: 150 billion and 200 billion barrels. He then made plausible estimates of future oil production rates for each of the two guesses. Even the more optimistic estimate, 200 billion barrels, led to a predicted peak of U.S. oil production in the early 1970s. The actual peak year turned out to be 1970.

Today, we can do something similar for world oil production. One educated guess of ultimate world recovery, 1.8 trillion barrels, comes from a 1997 country-by-country evaluation by Colin J. Campbell, an independent oil-industry consultant. In 1982, Hubbert's last published paper contained a world estimate of 2.1 trillion barrels. Hubbert's 1956 method leads to a peak year of 2001 for the trillion-barrel estimate and a peak year of 2003 or 2004 for 2.1 trillion barrels. The prediction based on 1.8 trillion barrels makes a better match to the most recent 10 years of world production.

"Hubbert's peak" confirmed

In 1962, I became concerned that the U.S. oil business might not be healthy by the time I was scheduled to retire. I was in no mood to move to Libya. My reaction was to get a photocopy of Hubbert's raw numbers; I made my own analysis using different mathematics. In my analysis, and in Hubbert's, the domestic oil industry would be down to half its peak size by 1998. Fortunately, universities were expanding rapidly . . . and I had no trouble moving into academe.

Hubbert's prediction was fully confirmed in the spring of 1971. The announcement was made publicly, but it was almost an encoded message. The *San Francisco Chronicle* contained this one-sentence item: "The Texas Railroad Commission announced a 100 percent allowable for next month." I went home and said, "Old Hubbert was right." It still strikes me as odd that understanding the newspaper item required knowing that the Texas Railroad Commission, many years earlier, had been assigned the task of matching oil production to demand. In essence, it was a government-sanctioned cartel. Texas oil production so dominated the industry that regulating each Texas oil well to a percentage of its capacity was enough to maintain oil prices. The Organization of Petroleum Exporting Countries (OPEC) was modeled after the Texas Railroad Commission. Just substitute Saudi Arabia for Texas. . . .

The actual peak year [of U.S. oil production] turned out to be 1970.

With Texas, and every other state, producing at full capacity from 1971 onward, the United States had no way to increase production in an emergency. During the first Middle East oil crisis in 1967 [when OPEC raised oil prices significantly], it was possible to open up the valves in Ward and Winkler Counties in west Texas and partially make up for lost imports. Since 1971, we have been dependent on OPEC.

After his prediction was confirmed, Hubbert became something of a folk hero for conservationists. In contrast to the hundreds of millions of years it took for the world's oil endowment to accumulate, most of the oil is being produced in 100 years. The short bump of oil exploitation on the geologic time line became known as "Hubbert's peak.". . .

Hubbert used oil production and oil reserves to predict the future. We scientists don't like to admit it, but we often guess at the answer and then gather up some numbers to support the guess. A certain level of honesty is required; if the numbers do not justify my guess, I don't fake the numbers. I generate another guess. Hubbert's oil prediction was just barely within the envelope of acceptable scientific methods. It was as much an inspired guess as it was hard-core science.

This cautionary note is needed here: in the late 1980s there were huge and abrupt increases in the announced oil reserves for several OPEC nations. Oil reserves are a vital ingredient in Hubbert's analysis. Earlier, each OPEC nation was assigned a share of the oil market based on the country's annual production capacity. OPEC changed the rule in the 1980s to consider also the oil reserves of each country. Most OPEC countries promptly increased their reserve estimates. These increases are not necessarily wrong; they are not necessarily fraudulent. "Reserves" exist in the eye of the beholder.

[Oil-producing countries will soon] market their remaining oil at mind-boggling prices.

Oil reserves are defined as future production, using existing technology, from wells that have already been drilled (not to be confused with the U.S. "strategic petroleum reserve," which is a storage facility for oil that has already been produced). Typically, young petroleum engineers unconsciously tend to underestimate reserves. It's a lot more fun to go into the boss's office next year and announce that there is actually a little *more* oil than last year's estimate. Engineers who have to downsize their previous reserve estimates are the first to leave in the next corporate downsizing.

The abrupt increase in announced OPEC reserves in the late 1980s was probably a mixture of updating old underestimates and some wishful thinking. A Hubbert prediction requires inserting some hard, cold reserve numbers into the calculation. The warm fuzzy numbers from OPEC probably give an overly optimistic view of future oil production. So who is supposed to know?

A firm in Geneva, Switzerland, called Petroconsultants, maintained a huge private database. One long-standing rumor said that the U.S. Central Intelligence Agency [CIA] was Petroconsultants' largest client. I would hope that between them, the CIA and Petroconsultants had inside information on the real OPEC reserves. This much is known: the loudest warnings about the predicted peak of world oil production came from Petroconsultants. My guess is that they were using data not available to the rest of us.

Ignoring the coming shortage

A permanent and irreversible decline in world oil production would have both economic and psychological effects. So who is paying attention? The news media tell us that increases in energy prices are caused by an assortment of regulations, taxes, and distribution problems. During the election campaign of 2000, none of the presidential candidates told us that the sky was about to fall. The public attention to the predicted oil shortfall is essentially zero.

In private, the OPEC oil ministers probably know about the articles in *Science, Nature,* and *Scientific American.* Detailed articles, with contrasting opinions, have been published frequently in the *Oil and Gas Journal.* Crude oil prices have doubled in 2000. I suspect that OPEC knows that a global oil shortage may be only a few years away. The OPEC countries can trickle out just enough oil to keep the world economies functioning until that glorious day when they can market their remaining oil at mind-boggling prices.

It is not clear whether the major oil companies are facing up to the problem. Most of them display a business-as-usual facade. My limited attempts at spying turned up nothing useful. A company taking the 2004–8 hypothesis seriously would be willing to pay top dollar for existing oil fields. There does not seem to be an orgy of reserve acquisitions in progress.

Internally, the oil industry has an unusual psychology. Exploring for oil is an inherently discouraging activity. Nine out of 10 exploration wells are dry holes. Only one in a hundred exploration wells discovers an important oil field. Darwinian selection is involved: only the incurable optimists stay. They tell each other stories about a Texas county that started with 30 dry holes yet the next well was a major discovery. "Never is heard a discouraging word." A permanent drop in world oil production beginning in this decade is definitely a discouraging word.

It looks as if an unprecedented [oil] crisis is just over the horizon.

Is there any way out? Is there some way the crisis could be averted?

New Technology. One of the responses in the 1980s was to ask for a double helping of new technology. Here is the problem: before 1995 (when the dot.com era began), the oil industry earned a higher rate of return on invested capital than any other industry. When oil companies tried to use some of their earnings to diversify, they discovered that everything else was less profitable than oil. Their only investment option was doing research to make their own exploration and production operations even more profitable. Billions of dollars went into petroleum technology development, and much of the work was successful. That makes it difficult to ask today for new technology. Most of those wheels have already been invented.

Drill Deeper. . . . There is an "oil window" that depends on subsurface temperatures. The rule of thumb says that temperatures 7,500 feet down are hot enough to "crack" organic-rich sediments into oil molecules. However, beyond 15,000 feet the rocks are so hot that the oil molecules are further cracked into natural gas. The range from 7,000 to 15,000 feet is called the "oil window." If you drill deeper than 15,000 feet, you can find natural gas but little oil. Drilling rigs capable of penetrating to 15,000 feet became available in 1938.

Drill Someplace New. Geologists have gone to the ends of the Earth in their search for oil. The only rock outcrops in the jungle are in the banks of rivers and streams: geologists waded up the streams picking leeches off their legs. A typical field geologist's comment about jungle, desert, or tundra was: "She's medium-tough country." As an example, at the very northernmost tip of Alaska, at Point Barrow, the United States set up Naval Petroleum Reserve #4 in 1923. As early as 1923, somebody knew that the Arctic Slope of Alaska would be a major oil producer.

Today, about the only promising petroleum province that remains unexplored is part of the South China Sea, where exploration has been delayed by a political problem. International law divides oil ownership at sea along lines halfway between the adjacent coastlines. A valid claim to an island in the ocean pushes the boundary out to halfway between the island and the farther coast. It apparently does not matter whether the island is just a protruding rock with every third wave washing over the rock. Ownership of that rock can confer title to billions of barrels of oil. You guessed it: several islands stick up in the middle of the South China Sea, and the drilling rights are claimed by six different countries. Although the South China Sea is an attractive prospect, there is little likelihood that it is another Middle East.

Speed Up Exploration. It takes a minimum of 10 years to go from a cold start on a new province to delivery of the first oil. One of the legendary oil finders, Hollis Hedberg, explained it in terms of "the story." When you start out in a new area, you want to know whether the oil is trapped in folds, in reefs, in sand lenses, or along faults. You want to know which are the good reservoir rocks and which are the good cap rocks. The answers to those questions are "the story." After you spend a few years in exploration work and drilling holes, you figure out "the story." For instance, the oil is in fossil patch reefs. Then pow, pow, pow—you bring in discovery after discovery in patch reefs. Even then, there are development wells to drill and pipelines to install. It works, but it takes 10 years. Nothing we initiate now will produce significant oil before the 2004–8 shortage begins.

Minimizing future disruptions

To summarize: it looks as if an unprecedented crisis is just over the horizon. There will be chaos in the oil industry, in governments, and in national economies. Even if governments and industries were to recognize the problems, it is too late to reverse the trend. Oil production is going to shrink. In an earlier, politically incorrect era the scene would be described as a "Chinese fire drill."

What will happen to the rest of us? In a sense, the oil crises of the 1970s and 1980s were a laboratory test. We were the lab rats in that experiment. Gasoline was rationed both by price and by the inconvenience of long lines at the gas stations. The increased price of gasoline and diesel fuel raised the cost of transporting food to the grocery store. We were told that 90 percent of an Iowa corn farmer's costs were, directly and indirectly, fossil fuel costs. As price rises rippled through the economy, there were many unpleasant disruptions.

Everyone was affected. One might guess that professors at Ivy League universities would be highly insulated from the rough-and-tumble world. I taught at Princeton from 1967 to 1997; faculty morale was at its lowest in the years around 1980. Inflation was raising the cost of living far faster than salaries increased. Many of us lived in university-owned apartments, and the university was raising our apartment rents in step with an imaginary outside "market" price. Our real standard of living went progressively lower for several years in a row. That was life (with tenure) inside the sheltered ivory tower; outside it was much tougher.

What should we do? Doing nothing is essentially betting against Hubbert. Ignoring the problem is equivalent to wagering that world oil production will continue to increase forever. My recommendation is for us to bet that the prediction is roughly correct. Planning for increased energy conservation and designing alternative energy sources should begin now to make good use of the few years before the crisis actually happens.

One possible stance, which I am not taking, says that we are despoiling the Earth, raping the resources, fouling the air, and that we should eat only organic food and ride bicycles. Guilt feelings will not prevent the chaos that threatens us. I ride a bicycle and walk a lot, but I confess that part of my motivation is the miserable parking situation in Princeton. Organic farming can feed only a small part of the world population; the global supply of cow dung [used as farming compost] is limited. A better civilization is not likely to arise spontaneously out of a pile of guilty consciences. We need to face the problem cheerfully and try to cope with it in a way that minimizes problems in the future. . . .

The experts' scenario for 2004–8 reads like the opening passage of a horror movie. You have to make up your own mind about whether to accept their scary account.

My own opinion is that the peak in world oil production may even occur before 2004. What happens if I am wrong? I would be delighted to be proved wrong. It would mean that we have a few additional years to reduce our consumption of crude oil. However, it would take a lot of unexpectedly good news to postpone the peak to 2010. My message would remain much the same: crude oil is much too valuable to be burned as a fuel.

2

World Oil Supplies Are Plentiful

Robert L. Bradley Jr.

Robert L. Bradley Jr. is president of the Institute for Energy Research in Houston, Texas, an organization that evaluates public policies in the oil, gas, and coal markets. He is also the author of Oil, Gas, and Government: The U.S. Experience.

Contrary to the predictions of some geologists and academics that the world is running out of oil, probable oil supplies should last over a hundred years. In addition, other fossil fuel sources—such as orimulsion, synthesized natural gas, and tar sand—show great promise when oil becomes more scarce and costly in the distant future. Human ingenuity will work to significantly limit the impact of any depletion of natural resources that may occur in the coming decades.

O nly a few years ago [mid-1990s] academics, businessmen, oilmen, and policymakers were almost uniformly of the opinion that the age of energy scarcity was upon us and that the depletion of fossil fuels was imminent. While some observers still cling to that view today, the intellectual tide has turned against doom and gloom on the energy front. Indeed, resource economists are almost uniformly of the opinion that fossil fuels will remain affordable in any reasonably foreseeable future.

Resources as far as the eye can see

Proven world reserves of oil, gas, and coal are officially estimated to be 45, 63, and 230 years of current consumption, respectively (Figure 1). Probable resources of oil, gas, and coal are officially forecast to be 114, 200, and 1,884 years of present usage, respectively.

Moreover, an array of unconventional fossil-fuel sources promises that, when crude oil, natural gas, and coal become scarcer (hence, more expensive) in the future, fossil-fuel substitutes may still be the best source fuels to fill the gap before synthetic substitutes come into play.

The most promising unconventional fossil fuel today is orimulsion, a tarlike substance that can be burned to make electricity or refined into petroleum. Orimulsion became the "fourth fossil fuel" [in addition to oil, coal, and natural gas] in the mid-1980s when technological improvements made Venezuela's reserves commercially exploitable. Venezuela's reserve equivalent of 1.2 trillion barrels of oil exceeds the world's known reserves of crude oil, and other countries' more modest supplies of the natural bitumen add to the total.

With economic and environmental (post-scrubbing) characteristics superior to those of fuel oil and coal when used for electricity generation, orimulsion is an attractive conversion opportunity for facilities located near waterways with convenient access to Venezuelan shipping. While political opposition (in Florida, in particular) has slowed the introduction of orimulsion in the United States, orimulsion has already penetrated markets in Denmark and Lithuania and, to a lesser extent, Germany and Italy. India could soon join that list. Marketing issues aside, this here-and-now fuel source represents an abundant backstop fuel at worst and a significant extension of the petroleum age at best.

The significance of orimulsion for the electricity-generation market may be matched by technological breakthroughs commercializing the conversion of natural gas to synthetic oil products. For remote gas fields, gas-to-liquids processing can replace the more expensive alternative of liquefaction. In mature markets with air quality concerns, such as in California, natural gas could become a key feedstock from which to distill the cleanest reformulated gasoline and reformulated diesel fuel yet. A half

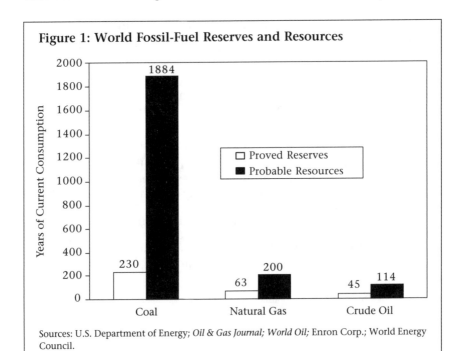

Figure 1: World Fossil-Fuel Reserves and Resources

Sources: U.S. Department of Energy; *Oil & Gas Journal; World Oil;* Enron Corp.; World Energy Council.

dozen competing technologies have been developed, several by oil majors who are committing substantial investments relative to government support. The widespread adaptation of gas-to-oil technologies could commercialize up to 40 percent of the world's natural gas fields that hitherto have been uneconomic.

Resource economists are almost uniformly of the opinion that fossil fuels will remain affordable.

In addition to orimulsion and synthesized natural gas, tar sand, shale oil, and various replenishable crops also have great promise, however uneconomic they now are, given today's technology and best practices. Michael Lynch of the Massachusetts Institute of Technology [MIT] estimates that more than 6 trillion barrels of potentially recoverable conventional oil and another 15 trillion barrels of unconventional oil (excluding coal liquefaction) are identifiable today, an estimate that moves the day of reckoning for petroleum centuries into the future.

The gas resource base is similarly loaded with potential interfuel substitutions, with advances in coal-bed methane and tight-sands gas showing immediate potential and synthetic substitutes from oil crops having long-run promise. If crude oil and natural gas are retired from the economic playing field, fossil fuels boast a strong "bench" of clean and abundant alternatives. Even the cautious Energy Information Administration of the U.S. Department of Energy conceded that "as technology brings the cost of producing an unconventional barrel of oil closer to that of a conventional barrel, it becomes reasonable to view oil as a viable energy source well into the twenty-second century."

Technological advances and increasing resources

Despite a century of doom and gloom about the imminent depletion of fossil-fuel reserves, fossil-fuel availability has been increasing even in the face of record consumption. World oil reserves today are more than 15 times greater than they were when record keeping began in 1948; world gas reserves are almost four times greater than they were 30 years ago; world coal reserves have risen 75 percent in the last 20 years. Thus, today's reserve and resource estimates should be considered a minimum, not a maximum. By the end of the forecast period, reserves could be the same or higher depending on technological developments, capital availability, public policies, and commodity price levels.

Technological advances continue to substantially improve finding rates and individual well productivity. Offshore drilling was once confined to fields several hundred feet below the ocean, for instance, but offshore drilling now reaches depths of several thousand feet. Designs are being considered for drilling beyond 12,000 feet.

Predictably, advances in production technology are driving down the cost of finding oil. In the early 1980s finding costs for new crude oil reserves averaged between $11.50 and $12.50 per barrel in the United States and most areas of the world. In the mid-1990s finding costs had fallen to around $7 per barrel despite 40 percent inflation in the interim. In the United States alone, finding costs dropped 40 percent between 1992 and 1996. That is perhaps the best indicator that oil is growing more abundant, not scarcer.

Finally, the amount of energy needed to produce a unit of economic goods or services has been declining more or less steadily. New technologies and incremental gains in production and consumption efficiency make the services performed by energy cheaper even if the original resource has grown more (or less) expensive in its own right.

Understanding resource abundance

How is the increasing abundance of fossil fuels squared with the obviously finite nature of those resources? "To explain the price of oil we must discard all assumptions of a fixed stock and an inevitable long-run rise and rule out nothing a priori," says M.A. Adelman of MIT. "Whether scarcity has been or is increasing is a question of fact. Development cost and reserve values are both measures of long-run scarcity. So is reserve value, which is driven by future revenues."

Natural resource economists have been unable to find a "depletion signal" in the data. A comprehensive search in 1984 by two economists at Resources for the Future found "gaps among theory, methodology, and data" that prevented a clear delineation between depletion and the "noise" of technological change, regulatory change, and entrepreneurial expectations. A more recent search for the depletion signal by [U.S. Department of Energy economist] Richard O'Neill et al. concluded:

> Care must be taken to avoid the seductiveness of conventional wisdom and wishful thinking. While the theory of exhaustible resources is seductive, the empirical evidence would be more like the bible story of the loaves and fishes. What matters is not exhaustible resource theories (true but practically dull) but getting supply to market (logistics) without disruption (geopolitics). While it is easy to see how political events may disrupt supply, it is hard to contrive an overall resource depletion effect on prices.

The facts, however, are explainable. Says Adelman:

> What we observe is the net result of two contrary forces: diminishing returns, as the industry moves from larger to smaller deposits and from better to poorer quality, versus increasing knowledge of science and technology generally, and of local government structures. So far, knowledge has won.

Human ingenuity and financial wherewithal, two key ingredients in the supply brew, are not finite but expansive. The most binding resource constraint on fossil fuels is the "petrotechnicals" needed to locate and extract the energy. Congruent with Adelman's theory, wages in the energy industry can be expected to increase over time, while real prices for energy can be expected to fall under market conditions. Under political conditions such as those that existed during the 1970s, however, the record of energy prices can be quite different.

There is no reason to believe that energy per se (as opposed to particular energy sources) will grow less abundant (more expensive) in our lifetimes or for future generations. "Energy," as Paul Ballonoff [author of *Energy: Ending the Never-Ending Crisis*] has concluded, "is simply another technological product whose economics are subject to the ordinary market effects of supply and demand." Thus, a negative externality cannot be assigned to today's fossil-fuel consumption to account for intergenerational "depletion." A better case can be made that a positive intergenerational externality is created, since today's base of knowledge and application subsidizes tomorrow's resource base and consumption.

[Fossil fuels like] tar sand, shale oil, and various replenishable crops . . . have great promise.

The implication for business decision-making and public policy analysis is that "depletable" is not an operative concept for the world oil market as it might be for an individual well, field, or geographical section. Like the economists' concept of "perfect competition," the concept of a nonrenewable resource is a heuristic, pedagogical device—an ideal type—not a principle that entrepreneurs can turn into profits and government officials can parlay into enlightened intervention. The time horizon is too short, and technological and economic change is too uncertain, discontinuous, and open-ended.

3

U.S. Energy Independence Is Not a Realistic Goal

Pietro S. Nivola

Pietro S. Nivola is the author of The Politics of Energy Conservation *and co-author of* The Extra Mile: Rethinking Energy Policy for Automotive Transportation. *He is also a senior fellow in the government studies program of the Brookings Institution, a nonpartisan think tank.*

The administration of President George W. Bush has made U.S. energy self-sufficiency a cornerstone of its energy policy. However, it is illogical to insist on producing oil in the United States at a greater cost than it can be produced elsewhere. The solution to protecting the U.S. economy against oil-supply disruptions from Middle East exporters is to develop the energy potential of America's neighbors, Canada and Mexico, through further liberalization of North American energy trade agreements. Such an energy strategy will preclude spending on wasteful domestic energy schemes—including producing synthetic fuels like ethanol—that will not improve America's energy security.

"Let us set as our national goal," President Richard Nixon proclaimed more than a quarter-century ago, "that by the end of this decade we will have developed the potential to meet our own energy needs without depending on any foreign sources."

Although President George W. Bush's energy plan is a lot less utopian, and indeed has certain merits, one of its suppositions, too, seems to be that Americans will be better off if they reduce "reliance on foreign energy." During the 2000 election campaign, the Bush camp rhapsodized about bygone days when "one of America's greatest strengths was its energy self-sufficiency." The theme has since been reiterated frequently, for the United States now produces at home less than half the oil it uses. In the words of the House majority whip, Tom DeLay, "America faces a serious degradation of our national security unless we move at once to reduce our dependence on foreign sources of energy." (Which "foreign sources of energy"? Presumably, we are not debating inflows from friendly places like Canada and Mexico. But hold on; I will return to that subject shortly.)

Pietro S. Nivola, "Energy Independence or Interdependence? Integrating the North American Energy Market," *Brookings Review*, vol. 20, Spring 2002, pp. 24–27. Copyright © 2002 by Brookings Institution. Reproduced by permission.

Cobbling at home

For all its persistent political fascination, the self-sufficiency logic is fundamentally flawed. In 1973 the United States imported little more than a third of the petroleum it consumed. Yet the U.S. economy then proved far more, not less, exposed to the shock of rising international oil prices than it was two years ago [2000] when those prices soared again while our dependence on foreign oil reached an all-time high. The extent of a nation's vulnerability, in other words, is not a function of how much energy it produces domestically or buys from abroad. Britain, for instance, produces more oil than it needs. But its self-sufficiency scarcely shielded British consumers from the sudden spike in gasoline prices in the summer of 2000. The reason is simple. Petroleum prices everywhere are set in a world market—and no country, even a net exporter, can readily repair to an energy policy that says, in effect, "Stop the world. I want to get off."

More consequential for economic stability than the share of fuel supplied by foreign sources is a nation's relative energy-intensity and susceptibility to inflationary pressures. To produce a dollar of GDP [gross domestic product] America today requires about 30 percent less energy than it did 30 years ago. So our economy now is less vulnerable to the effects of energy price increases. The economy is also far less inflation prone than it was in the 1970s. Consequently, even the tripling of global crude-oil prices between 1998 and 2000 did little damage.

A nation's vulnerability [to rising oil prices] . . . is not a function of how much energy it produces domestically or buys from abroad.

Although you might not know it from listening to the rhetoric in Washington, this country still produces all but about a quarter of the "energy" it needs. Yes, imports of oil have increased (mostly because Americans choose to drive far more, and use much less efficient motor vehicles, than do consumers in other industrial countries), but imported oil is just one part of the picture. Nearly all of what propels the nation's electric generators—coal, gas, nuclear power, hydropower, and non-hydro renewables—is domestically produced. About 85 percent of our primary heating fuel, natural gas, is made in the U.S.A. (almost all the rest comes from Canada).

The quest for ever-greater energy independence was fanciful in Nixon's time and would be no less quixotic today. The days when the United States was a low-cost producer of oil are history. Squeezing additional supplies from the remaining domestic reserves will require increasingly elaborate operations, many of which may be prohibitively expensive if proper environmental safeguards are factored in and if world prices sag, as they have repeatedly in the past 15 years. In a free market, it is hard to see how such projects can compete with various foreign sources where production costs are lower. Most of the oil extracted by the Organization of Petroleum Exporting Countries (OPEC) costs less than $5 a barrel to produce.

It is wasteful to insist on making at home those commodities that can usually be supplied more cheaply from somewhere else—just as it makes no sense for me to cobble my own shoes instead of "importing" them from a shoe store. Funneling scarce resources into enterprises in which we no longer have a comparative advantage means leaving less capital and labor available to other industries—ones that could put those resources to better use. Society's living standards are lowered, not secured, by the pursuit of autonomy.

Prudent policy

All this is not to say that no additional reserves of oil are worth recovering within U.S. territory. Wherever feasible, they should be found and used. Adding their output to the world's supply is desirable, just as it is worthwhile to get more from any industry that can competitively make products the world needs.

Nor do I mean to suggest that a disruption of the global oil supply would no longer have an adverse impact on the U.S. economy. Suppose that another crisis in the Middle East were to take a large volume of oil—say, 7 million barrels a day (mbd)—off the market. Suppose, also, that after drawing down perhaps 2.5 mbd, from strategic petroleum reserves no other sources were available to make up for the shortage. Brookings economist George L. Perry estimates that the world price of crude could hit $75 a barrel. An increase that big could add as much as 5 percentage points to inflation and would promptly deepen the recession throughout the industrialized world.

Clearly, building up emergency stockpiles in the industrial countries is the first line of defense against such a disruption. Having access to spare production capacity outside the disrupted fields is also helpful. If the strategic reserves contribute 2.5 mbd and the extra capacity could add 1 mbd, a potential 7 mbd loss would be reduced by half. Under these circumstances, according to Perry, the price of oil would go to only $32 a barrel, so that regular gasoline would cost roughly $1.75 a gallon—a change well within the range of price variation in recent years. But notice: the surge capacity is a global, not a national, imperative. As long as it is available somewhere amid the earth's extensive assortment of reliable producing areas, much of the global shortfall (and the price shock for American consumers) will be offset.

It is wasteful to insist on making at home those commodities [such as oil] that can usually be supplied more cheaply from somewhere else.

Some of this cushion already exists. The supply side of the international petroleum market today is a lot more diversified than it was in the dark days of 1973, when the Arab oil embargo wrought havoc. The share of oil the United States buys from the Persian Gulf now represents less than a quarter of U.S. oil imports, down from almost 30 percent in 1980. Fully half of the oil we import comes from within the Western hemisphere. A number of large suppliers, outside OPEC as well as within it, are not operating at full capacity. If past is prelude, they would almost certainly put more oil on line amid the high prices of a tight market.

To gain additional insurance in an economically efficient fashion over the longer haul, policymakers should enhance trade relations with the stable suppliers that have abundant energy resources. Conveniently, at least two of them happen to be in the immediate vicinity. President Bush, to his credit, has spoken of the "need to recognize the energy potential of our neighbors, Canada and Mexico."

Opportunities in North America

The United States is the world's second largest producer of oil, Mexico the eighth. But with proved reserves around 23 percent larger than those of the United States, Mexico holds considerable promise for increased production. The same goes for our northern neighbor. No country, not even Saudi Arabia, exports more oil to the United States than does Canada. In the years ahead, Canada is likely to make a larger contribution as much more output from recent discoveries, such as Newfoundland's enormous Hibernia Field, comes on stream. The bulk of Canada's hydrocarbon resource base is still unrecoverable at current prices and with existing technology, but it is potentially immense. Alberta's oil sands, for instance, might contain the equivalent of an estimated 300 billion barrels of oil, more than all of Saudi Arabia's proven reserves. To put matters in perspective, if eventually even just one-quarter of this source were to become available, it would meet all the combined demand of both countries for nearly a decade (at current rates of consumption).

Policymakers should enhance trade relations with the stable suppliers that have abundant energy resources.

Opportunities for expanded development and trade of natural gas throughout North America ought to be more fully exploited as well. Canada now furnishes around 14 percent of U.S. requirements. New fields are being developed off Canada's east coast between Newfoundland and Nova Scotia. In addition to providing heating fuel for Nova Scotia and New Brunswick, the offshore source could at last bring an end to New England's anomalous dependence on heating oil.

More could be done. Vast quantities of gas, possibly several hundred trillion cubic feet, remain untapped in an area stretching across the Western Canada Sedimentary Basin to Alaska. Known deposits in Alaska and the Yukon alone exceed 70 trillion cubic feet—more than three years of total U.S. consumption at present levels. Official estimates of Mexican proven reserves of natural gas are reportedly half that, but recent estimates by the Gas Technology Institute suggest that as much as 65 trillion cubic feet of gas could turn up in the Burgos Basin of northeastern Mexico.

The importance of bringing more natural gas to the North American market cannot be overstated. Gas can readily displace oil in various ways—as a heating fuel, as a fuel for electric generators (substituting for the oil-fired ones that comprise about half of Mexico's generating capacity, for example), and to an extent, even as a transportation fuel. More important, this relatively clean-burning fuel is the leading alternative to coal. To meet America's anticipated demand for electricity in the near future, as much as 200,000 additional megawatts of electric power may have to be produced nationwide. Across the continent, power plants will spew fewer air pollutants, particularly greenhouse gas emissions, if they are powered by natural gas instead of coal.

NAFTA's unfinished business

The energy markets of the United States, Canada, and Mexico would have realized more of their potential by now if the North American Free Trade Agreement (NAFTA) had integrated them more completely. But in the energy sector, NAFTA stopped short of achieving a truly open framework for trilateral trade and investment.

The United States is not likely to be well served by any national energy strategy that favors . . . a buy-American bias.

The main roadblock remains on the Mexican side, where constitutional provisions, nationalistic sentiment, and raw politics combine to maintain state-owned monopolies that retard growth of that country's oil, gas, and electricity industries. According to the World Bank's estimates, Mexico ought to be financing approximately $10 billion each year in energy development and infrastructure just to meet its own projected demand, never mind those of its trading partners. Yet, Petróleos Méxicanos (Pemex) is required by law to transfer as much as a third of its annual revenues to the Mexican treasury. During 2000, close to $30 billion in Pemex earnings was turned over, funding 37 percent of the government's general budget but leaving the company unable to plow adequate revenues back into new exploration and production. Not only do Pemex and the public power company, Comisión Federal de Electricidad (CFE), perennially underinvest: the system of national enterprises (and, in the case of CFE, underpriced electrical rates) also discourages private investors from underwriting Mexico's essential energy projects.

Most of the liberalization of energy trade has been between Canada and the United States. By the time NAFTA was ratified in 1993, Canada had jettisoned the statist energy regime it had erected some 15 years earlier. The government-run monopoly, Petro-Canada, was largely privatized. Gone as well were restrictions on foreign ownership and obstructive price regulations, including the mercantilist practice of imposing supra-market export prices. The result was a rapid rise in Canadian exports after 1983, especially to parts of the American West. By 1991 Canadian natural gas was satisfying approximately 25 percent of California's gas consumption.

Canada, unlike Mexico, did not insist on major exclusivity provisions for energy industries under NAFTA. Thus U.S. and Canadian interests have launched important cooperative ventures in recent years. The Alliance Pipeline—extending over 2,300 miles from western Canada to Chicago—is probably the most striking example. The jointly financed line can carry 1.4 billion cubic feet of gas to the Midwest, and connecting links will eventually deliver a substantial share to locations as far away as Pennsylvania and New York.

U.S. [energy] policy should seek deeper integration with [Canada and Mexico].

But U.S.-Canada energy flows still are not entirely unfettered. Ottawa can negotiate international trade treaties, but jurisdiction over the disposition of natural resources rests with the provinces. Local disputes over key decisions, such as the routing of natural gas lines from the Arctic to the continental United States, continue to cause delays.

All three NAFTA countries could save energy and score big environmental gains if their electric utility grids interfaced seamlessly. Fewer additional generating facilities might be necessary in the future if power could be readily wheeled around the continent to wherever it is needed. But inadequate transmission capacity, unsynchronized connections, and insufficient deregulation (open access) in various local systems continue to pose obstacles. The deficiencies, moreover, are not limited to links between the United States and Mexico, where cross-border sales of electricity are still negligible. To a degree, problems also persist in and among U.S. regions and some Canadian provinces.

What next?

The United States is not likely to be well served by any national energy strategy that favors, in effect, a buy-American bias when international trade can meet a sizable share of our energy requirements more affordably. Recent events, if anything, underscore this conclusion.

To succeed without massive subsidies, government policies to encourage homemade energy would have to be supported by a sustained period of steep market prices and also by a long suspension of politics as usual. Neither is likely. The ink was barely dry on the Bush administration's National Energy Policy [introduced by Vice President Dick Cheney in May 2001] when market prices shifted unexpectedly. Natural gas, which had run to $10 per 1,000 cubic feet in early 2001, was closer to $3 by [the summer of 2001] and headed lower. Crude oil plunged from about $30 a barrel in early September 2001 to around $17 a barrel by mid-November. Everywhere, including California (where spot prices of electricity had briefly soared), the energy "crisis"—and thus the urgency of having a "plan"—had faded.

In any case, even the best-laid energy plans are perforce politicized. The Bush initiative is no exception. In the name of furthering "energy independence," a multitude of interest groups make a pitch for their pet projects. Never mind how costly, amid the legislative logrolling any homespun energy scheme—from ethanol farms to synthetic fuels—can make headway. By the time the House of Representatives had finished with the president's proposals, for example, it had authorized more than $5 billion in subventions to promote an oxymoron: "clean coal." (Power plants using coal treated by so-called fluidized bed technology still emit about 10 times more smog-causing nitrogen oxide than do comparably sized plants fueled by natural gas.)

Instead of futile planning to quarantine "foreign energy providers," U.S. policy should seek deeper integration with some of them, especially the ones next door. In accordance with the aspirations of NAFTA—a genuinely open market for commerce and investment—the three member countries need to be shedding more of their anachronistic energy regulations. In particular, Mexico should be urged to privatize Pemex and the CFE. Before his election President Vicente Fox had intimated that the Mexican government ought to reduce its reliance on Pemex as a cash cow. One way for the United States to encourage reform south of the border might be to swap liberalized immigration rules up here for a firm commitment to restructuring (or, at a bare minimum, much lighter taxation) of the state-controlled energy industries down there.

The United States and Canada, too, can take additional steps congruent with the free-market principles of NAFTA. The installation of new power lines linking U.S. and Canadian utilities, for example, will lag as long as regulators on both sides of the border prevent the utilities from profiting fully from these investments. And negotiating other vital regulatory reforms with our trading partners is complicated by the perpetuation of what they deem to be U.S. trade barriers. Some Canadian provinces complain that "mandated renewable portfolios" in many U.S. states (which allocate substantial market shares of electricity sales to U.S. producers of renewable energy) interfere with exports of electric power from Canada. In 1999, a coalition of independent U.S. oil producers asked the Commerce Department to impose anti-dumping duties on Mexican crude oil.

If politicians in Washington are serious about improving America's energy security, they would do well to clear away such hindrances, and in Mr. Bush's words, "make it easier for buyers and sellers of energy to do business across our borders."

4

The United States Invaded Iraq to Control Its Oil Reserves

Michael T. Klare

Michael T. Klare is the author of Resource Wars: The New Landscape of Global Conflict *and a professor of peace and world security studies at Hampshire College in Amherst, Massachusetts.*

President George W. Bush justified the U.S.-Britain invasion of Iraq in March 2003 by asserting that Iraqi president Saddam Hussein was illegally maintaining an arsenal of weapons of mass destruction, including chemical and biological weapons. In addition, the American public was told that Hussein was funding international terrorist organizations. Other countries, however, such as North Korea and Pakistan, pose a greater threat to U.S. security, and there is no evidence that Hussein funded terrorists. The real impetus behind the invasion of Iraq is U.S. dependence on foreign oil: The Bush administration views unrestricted access to Iraq's enormous oil reserves as vital to maintaining America's position as a world superpower.

Editor's note: This article was written shortly before the U.S.-Britain invasion of Iraq in March 2003, which led to the overthrow of Iraqi president Saddam Hussein. Hussein had been president since 1979.

The United States is about to go to war with Iraq. By early February 2003, 100,000 United States troops had already been deployed near Iraq, and another 75,000 or so were on their way to the region. Although most European leaders express satisfaction with the UN [United Nations] weapons inspections process, Secretary of Defense Donald Rumsfeld and other top administration officials have indicated that they will never be satisfied by inspections—only the voluntary disclosure by Iraq of prohibited weapons said to be in its possession by Washington will convince

Michael T. Klare, "For Oil and Empire? Rethinking War with Iraq," *Current History*, vol. 102, March 2003, pp. 129–35. Copyright © 2003 by Current History, Inc. Reproduced by permission.

them and the president [George W. Bush]. War, it appears, is inevitable unless Saddam Hussein is overthrown by members of the Iraqi military or is persuaded to abdicate his position and flee Baghdad [Iraq's capital], leaving the country in the hands of people willing to do Washington's bidding. . . .

Pointing in the wrong direction

In their public pronouncements, President Bush and his associates have advanced three reasons for going to war with Iraq and ousting Saddam Hussein: 1) to eliminate Saddam's WMD [weapons of mass destruction] arsenals; 2) to diminish the threat of international terrorism; and 3) to promote democracy in Iraq and the surrounding areas.

These are powerful motives for war. But are they genuine? Are they what is actually driving the rush to war? To answer this, we need to examine each motive in turn. In doing so, it is necessary to keep in mind that the United States cannot erase all the world's threats. If the United States commits hundreds of thousands of American troops and tens or hundreds of billions of dollars to the conquest, occupation, and reconstruction of Iraq, it cannot easily do the same in other countries; the United States simply does not have the resources to invade and occupy every country that poses a hypothetical threat to the United States or deserves regime change. A decision to attack Iraq means a decision to refrain from other actions that might also be important for American security or the good of the world.

Reducing the risk of a WMD attack on the United States is the reason most often given by the administration for going to war with Iraq. A significant WMD attack on the United States would be a terrible disaster, and it is appropriate for the president to take effective and vigorous action to prevent this from happening. If this is, in fact, Bush's primary concern, then he should pay the closest attention to the greatest threat of WMD usage against the United States, and deploy available United States resources—troops, dollars, and diplomacy—accordingly. But is this what Bush is actually doing? The answer is no. Anyone who takes the trouble to examine the global WMD proliferation threat closely and to gauge the relative likelihood of various WMD scenarios would have to conclude that the greatest threat of WMD usage against the United States at present comes from North Korea and Pakistan, not Iraq.

The greatest threat of WMD [weapons of mass destruction] usage against the United States . . . comes from North Korea and Pakistan, not Iraq.

North Korea and Pakistan pose greater WMD threats for several reasons. Both possess much larger WMD arsenals than Iraq. Pakistan maintains several dozen nuclear warheads along with missiles and planes capable of delivering them hundreds of miles; it is also suspected of having developed chemical weapons. North Korea is thought to possess sufficient plutonium to produce one or two nuclear devices along with the capacity to manufacture several more; it also has a large chemical weapons stockpile and a formidable array of ballistic missiles. Iraq, by contrast, has no nuclear weapons today and is thought to be several years away from producing one, even under the best of circumstances. Iraq may have some chemical and biological weapons and a dozen or so Scud-type missiles that were hidden at the end of the 1991 Persian Gulf War, but it is not known whether any of these items are still in working order and available for military use. . . .

The Bush administration has also indicated that war with Iraq is justified to prevent Iraq from providing WMD to anti-American terrorists. The transfer of WMD technology to terrorist groups is a genuine concern—but Pakistan is where the greatest threat of a transfer exists, not Iraq. In Pakistan many senior military officers are known to harbor sympathy for Kashmiri militants and other extremist Islamic movements; with anti-Americanism intensifying throughout the region, it is possible that these officers could provide militants with some of Pakistan's WMD and technology. The current leadership in Iraq has no such ties with Islamic extremists; on the contrary, Saddam has been a life-long enemy of the militant Islamists and they generally view him in an equally hostile manner.

A United States effort to oust Saddam Hussein . . .
will not diminish the wrath of Islamic [terrorists].

It follows from this that a policy aimed at protecting the United States from WMD attacks would identify Pakistan and North Korea as the leading concerns and put Iraq in a rather distant third place. But this is not, of course, what the administration is doing. Instead, it has minimized the threat from Pakistan and North Korea and focused almost exclusively on the threat from Iraq. Protecting the United States from WMD attack is not the primary justification for invading Iraq; if it were, the discussion would be centered on undertaking an assault on Pakistan or North Korea, not Iraq.

Fuel for the fire

The administration has argued at great length that an invasion of Iraq and the ouster of Saddam Hussein would constitute the culmination of and the greatest success in the war against terrorism. Why this is so has never been made entirely clear, but it is said that Saddam's hostility toward the United States somehow sustains and invigorates the terrorist threat to this country. It follows, therefore, that the elimination of Saddam would result in a great defeat for international terrorism and decisively weaken its capacity to attack the United States.

Were any of this true, an invasion of Iraq might make sense from an antiterrorism point of view. But there simply is no evidence of this; if anything, the opposite is true. From what we know of Al Qaeda[1] and similar

---

1. Islamic fundamentalist terrorist organization responsible for the September 11, 2001, attacks against the United States

organizations, the objective of Islamic extremists is to overthrow any government in the Muslim world that does not adhere to a fundamentalist version of Islam and replace it with one that does. Under Al Qaeda doctrine, the secular Baathist regime in Iraq must be swept away, along with the equally deficient governments in Egypt, Jordan, and Saudi Arabia. A United States effort to oust Saddam Hussein and replace his regime with another secular government—this one kept in place by American military power—will not diminish the wrath of Islamic extremists, but fuel it.

Reason for skepticism about the . . . Bush administration's commitment to democracy in the Middle East . . . stems from [its] close relationships with other [dictators].

In addressing this matter, moreover, it is necessary to keep the Israeli-Palestinian struggle in mind. For most Arab Muslims, whatever their views of Saddam Hussein, the United States is a hypocritical power because it tolerates (or even supports) the use of state terror by Israel against the Palestinians while it makes war against Baghdad for carrying out brutal acts against the Iraqi people. This perception fuels the anti-American current that runs through the Muslim world. An American invasion of Iraq will not quiet that current, but excite it. It is exceedingly difficult to see how a United States invasion of Iraq will produce a stunning victory in the war against terrorism; if anything, it will trigger a new round of anti-American violence. This makes it difficult to conclude that the administration is motivated by antiterrorism in seeking to topple Hussein.

Making Iraq safe for democracy

The removal of Saddam Hussein, it is claimed, will also clear a space for the Iraqi people (under American guidance, of course) to establish a truly democratic government and serve as a beacon and inspiration for the spread of democracy throughout the Islamic world. Certainly, the spread of democracy to the Islamic world would be a good thing, and should be encouraged. But is there any reason to believe that the administration is motivated by a desire to spread democracy in its rush to war with Iraq?

History sows doubt. Many of the top leaders of the current administration, particularly Secretary of Defense Rumsfeld and Vice President Dick Cheney, embraced Saddam Hussein's dictatorship in the 1980s when Iraq was the "enemy of our enemy" (that is, Iran) and was thus considered a de facto friend. Under its "tilt" toward Iraq, the Reagan administration decided to assist Iraq in its war against Iran during the Iran-Iraq War in the 1980s. As part of this policy, President Ronald Reagan removed Iraq from the list of countries that supported terrorism, thus permitting the provision of billions of dollars' worth of agricultural credits and other forms of assistance to Hussein. The bearer of this good news was none other than Donald Rumsfeld, who traveled to Baghdad and met with Hussein in December 1983 as a special representative of President Reagan.

At the same time, the Department of Defense, then headed by Caspar

Weinberger, provided Iraq with secret satellite data on Iranian military positions. This information was provided to Saddam even though United States leaders were informed by a senior State Department official on November 1, 1983 that the Iraqis were using chemical weapons against the Iranians "almost daily"; they were also aware that United States satellite data could be used by Baghdad to pinpoint chemical weapons attacks on Iranian positions. Cheney, who succeeded Weinberger as secretary of defense in 1989, continued the practice of supplying Iraq with secret intelligence data. Not once did Rumsfeld or Cheney speak out against Iraqi chemical warfare use or suggest that the United States discontinue its support of the Hussein dictatorship during this period. The current leadership cannot claim a principled objection to dictatorial rule in Iraq: it is only when Saddam is threatening the United States instead of America's enemies that it cares about his tyrannical behaviors.

Since World War II, . . . it has been American policy to ensure . . . unrestrained access to [oil from] the Persian Gulf.

Reason for skepticism about the current Bush administration's commitment to democracy in the Middle East also stems from the administration's close relationships with a number of other dictatorial or authoritarian regimes in the region. Most notably, the United States has developed ties with the post-Soviet dictatorships in Azerbaijan, Kazakhstan, and Uzbekistan since the war in Afghanistan.[2] Each of these countries is ruled by a Stalinist[3] dictator who once served as a loyal agent of the Soviet empire. Heydar Aliyev in Azerbaijan, Nursultan Nazarbayev in Kazakhstan, and Islam Karimov in Uzbekistan. Only slightly less odious than Saddam Hussein, these tyrants have been welcomed to the White House and showered with American aid and support. Kuwait and Saudi Arabia, two of America's other close regional allies, are also not even remotely democratic. It is difficult to accept the argument that the Bush administration is motivated by a love of democracy in wanting to oust Saddam when it has been so quick to embrace patently undemocratic regimes that have agreed to do its bidding.

Oil and empire

If concerns about weapons of mass destruction, terrorism, and the export of democracy do not explain the administration's determination to oust Saddam Hussein, what does? The answer is a combination of three factors, all related to the pursuit of oil and the preservation of America's status as the paramount world power.

Since the end of the cold war, policymakers in the United States (whether Democratic or Republican) have sought to preserve America's sole superpower status and prevent the rise of a "peer competitor" that

---

2. A U.S.-led invasion was conducted shortly after the September 11, 2001, terrorist attacks.   3. a reference to Joseph Stalin, leader of the Soviet Union from 1924 to 1953

could challenge its paramount position. At the same time, American leaders have become increasingly concerned about the country's growing dependence on imported oil—especially oil from the Persian Gulf. The United States now relies on foreign oil for 55 percent of its energy requirements, and this is expected to rise to 65 percent in 2020 and continue to grow thereafter. This dependency is the Achilles' heel of American power: unless Persian Gulf oil is kept under American control, the ability of the United States to remain the dominant world power would be put into question.

These concerns undergird the three real motives for a United States invasion of Iraq. The first derives from America's dependence on Gulf oil and from the principle formally enshrined in the Carter Doctrine, that the United States will not permit a hostile state to achieve a position that allows it to threaten America's access to the Gulf. The second is the pivotal role played by the Persian Gulf in supplying oil to the rest of the world: whoever controls the Gulf automatically maintains a stranglehold on the global economy; the Bush administration wants that power to be the United States. The third factor is anxiety about the future availability of oil: the United States has become increasingly dependent on Saudi Arabia, and Washington is desperate to find an alternative to the Saudis if access to that country is curtailed. The only nation in the world with reserves to compensate for the loss of Saudi Arabia is Iraq.

Control over the flow of Persian Gulf oil is . . . consistent with the [Bush] administration's declared goal of attaining . . . military superiority over all other nations.

Since World War II, when American policymakers first acknowledged that the United States would someday become dependent on Middle Eastern oil, it has been American policy to ensure that the United States will always have unrestrained access to the Persian Gulf. At first, the United States relied on Britain to protect American access to the Gulf, and then, when Britain pulled out of the area in 1971, the United States chose to rely on the Shah of Iran. But when, in 1979, the shah was overthrown by Islamic militants loyal to the Ayatollah Ruhollah Khomeini, the administration of President Jimmy Carter decided that the United States would have to assume responsibility to protect the flow of oil. The result was the policy now known as the Carter Doctrine. Unrestricted access to the Persian Gulf is a vital interest of the United States, Carter affirmed in his 1980 State of the Union address; in protection of that interest, the United States would employ "any means necessary, including military force."

This principle was first invoked by President Reagan in 1987, during the Iran-Iraq War, when Iranian gunboats fired on Kuwaiti oil tankers and the United States navy began escorting Kuwaiti tankers through the Gulf. President George H. W. Bush invoked it again in August 1990, when Iraq invaded Kuwait and posed an implied threat to Saudi Arabia. Bush the elder responded to that threat by driving the Iraqis out of Kuwait in Operation Desert Storm; he did not, however, continue the war and remove

Saddam Hussein. Instead, the United States engaged in the "containment" of Iraq, which entailed the maintenance of an air and sea blockade of that country.

President Bush the younger now wants to abandon containment and "finish" Operation Desert Storm. The underlying principle for military action is still the Carter Doctrine. Iraq under Saddam is an implied threat to United States access to Persian Gulf oil, and so the Iraqi leader must be removed. Vice President Cheney noted as much in his August 26, 2002 speech before an audience of the Veterans of Foreign Wars [VFW]:

> Armed with an arsenal of these weapons of terror and seated atop 10 percent of the world's oil reserves, Saddam Hussein could then be expected to seek domination of the entire Middle East, take control of a great portion of the world's energy supplies, directly threaten America's friends throughout the region, and subject the United States or any other nation to nuclear blackmail.

This is, in essence, a direct invocation of the Carter Doctrine.

Attaining economic and military superiority

Cheney's VFW speech echoes comments he made 12 years earlier before the Senate Armed Services Committee after the Iraqi invasion of Kuwait:

> Iraq controlled 10 percent of the world's reserves prior to the invasion of Kuwait. Once Saddam Hussein took Kuwait, he doubled that to approximately 20 percent of the world's known oil reserves. . . . Once he acquired Kuwait and deployed an army as large as the one he possesses [on the border of Saudi Arabia], he was clearly in a position to dictate the future of worldwide energy policy, and that gave him a stranglehold on our economy and on that of most of the other nations of the world as well.

The language of Cheney's 1990 testimony also drives the second administration objective in overthrowing Hussein: whoever controls the flow of Persian Gulf oil has a "stranglehold" not only on the American economy but also on the economies of "the other nations of the world as well." This is a powerful image, and captures perfectly the administration's thinking about the Gulf, except in reverse: by serving as the region's dominant power, the United States maintains a "stranglehold" over the economies of other nations. This gives the United States extraordinary leverage in world affairs, and explains to some degree why countries like Japan, Britain, France, and Germany—which are even more dependent on Persian Gulf oil than the United States—ultimately defer to Washington on major international issues (such as Iraq) even when they disagree with it.

Maintaining control over the flow of Persian Gulf oil is also consistent with the administration's declared goal of attaining permanent military superiority over all other nations. A single theme stands out in administration statements on United States national security policy: the United States must prevent any potential rival from ever reaching the point where it could compete with the United States on equal standing. As presented in *The National Security Strategy of the United States of America* (released by the administration in September 2002), this principle holds that American forces must be "strong enough to dissuade potential adversaries from pursuing a military build-up in hopes of surpassing, or equaling, the power of the United States."

There is only one way to . . . reduce . . . reliance on [oil from] Saudi Arabia: by taking over Iraq and using it as an alternative oil supplier.

One way to accomplish this, of course, is to pursue advances in technology that allow the United States to remain ahead of all potential rivals in military systems—which is what the administration hopes to accomplish by adding hundreds of billions of dollars to the Department of Defense budget through 2008. Another way to do this is to maintain an oil stranglehold on the economies of potential rivals so that they will refrain from challenging the United States out of fear of being choked to death through the denial of vital energy supplies. Japan and the European countries are already in this vulnerable position, and will remain so for the foreseeable future; but now China is also moving in this direction as it becomes increasingly dependent on oil from the Persian Gulf. Like the United States, China is running out of oil, and like the United States, it has nowhere to go to make up the difference except the Gulf. But since the United States controls access to the Gulf, and China lacks the power to break that grip, the United States can keep China in a vulnerable position indefinitely. The removal of Saddam Hussein and his replacement by someone beholden to the United States is a key part of a broader United States strategy aimed at assuring permanent American global dominance. Or, as Harvard's Michael Ignatieff put it in his seminal January 5, 2003 *New York Times Magazine* essay on America's emerging empire, the concentration of so much oil in the Gulf "makes it what a military strategist would call the empire's center of gravity."

Iraq: The alternative to Saudi Arabian oil

Finally, there is the issue of America's long-term energy dilemma. The United States uses oil to supply about 40 percent of its energy needs. At one time it relied almost entirely on domestic oil, but the demand for oil has continued to grow and America's domestic fields—among the oldest in the world—are gradually being exhausted. The need for imported oil will thus grow with each passing year. Most of the world's untapped oil—at least two-thirds of it—is located in the Persian Gulf. The United States can rip up Alaska and extract every drop of oil in the state, but that would reduce its dependence on imported oil by only about 2 to 3 percent—an insignificant amount. It could also rely for a share of its oil needs on non-Gulf suppliers such as Russia, Venezuela, the Caspian Sea states, and the nations of Africa, but their oil reserves are smaller than those of the Persian Gulf countries and are being extracted at a much quicker rate. The further you look into the future, the greater America's dependence on the Gulf becomes.

At present America's reliance on Persian Gulf oil means dependence on Saudi Arabia, which has more oil than any other country—about 250 billion barrels, or one-fourth of world reserves. This gives Saudi Arabia considerable indirect influence over the United States economy and the American way of life. Saudi Arabia is also a major power in the Organization of Petroleum Exporting Countries [OPEC][4] and can control the global price and supply of oil. This makes American officials nervous, especially when the Saudis can use their power to put pressure on the United States to alter its policies in other areas, such as the Israeli-Palestinian conflict.

American leaders would thus like to reduce the country's dependence on Saudi Arabia. But there is only one way to permanently reduce America's reliance on Saudi Arabia: by taking over Iraq and using it as an alternative oil supplier. Iraq possesses 112 billion barrels in proven oil reserves, and as much as 200 billion to 300 billion barrels of potential reserves. By occupying Iraq and installing a government friendly to it, the United States will solve its long-term oil-dependency dilemma for a decade or more. And that is a major consideration in the administration's policy choices regarding Iraq.

This set of factors explains the Bush administration's determination to go to war with Iraq—not concern over WMD, not terrorism, not the spread of democracy. But do these objectives—access to and control over critical oil supplies—justify a war on Iraq? Some Americans may think so. There are, indeed, advantages to being positioned to control the world's second-largest source of untapped petroleum. American motorists will be able to afford the gas to fuel their SUVs, vans, and pickup trucks for another decade, and maybe longer. There will also be many more jobs in the military and in the military-industrial complex, or as representatives of American multinational corporations (although, with respect to the latter, I would not advise traveling in most of the rest of the world unless accompanied by a small army of bodyguards).

But there will also be a price to pay. Empires tend to require the militarization of society, and that will mean increased spending on war and reduced spending on education and other domestic needs. It will also entail more secrecy and government intrusion into the private lives of American citizens. All this has to be entered into the equation. And the answer to the question has to be no: the construction and maintenance of empire are not worth the price.

---

4. a cartel of oil-producing nations

5

The United States Did Not Invade Iraq to Control Its Oil Reserves

Jerry Taylor

Jerry Taylor is director of natural resource studies at the Cato Institute, a libertarian think tank based in Washington, D.C.

The contention of peace activists that the United States and Britain invaded Iraq in March 2003 to control its large oil reserves is based on misleading arguments. If the war was fought simply to ensure greater U.S. access to Iraqi oil, then President George W. Bush could have lifted the trade embargo imposed on Iraqi exports in 1991 for far less cost and political risk. Moreover, the fact that the U.S. government has promised Iraq that it would maintain control over its own lucrative oil contracts once the war was over shows that the United States is not going to war to control Iraq's oil reserves. President Bush was justified in leading an invasion of Iraq because its oil reserves generate huge profits, which, in the hands of a rogue leader like the now-deposed Iraqi president Saddam Hussein, could fund a dangerous military arsenal.

Editor's note: This article was written one week prior to the U.S.-Britain invasion of Iraq in March 2003.

Is the coming war with Iraq about oil when all is said and done? The anti-war movement seems to think so. I am not so sure.

Addressing an intolerable danger

Unless the peace movement has discovered telepathy, I doubt that it's in any better position to divine the hidden thoughts or secret motivations of [President] George W. Bush and [British prime minister] Tony Blair than I am. Arguing about unstated motives, therefore, is a waste of time—

claims cannot be proven or disproven.

Is it so difficult, however, to imagine that both Bush and Blair sincerely believe—rightly or wrongly—that a well-armed Iraq poses an intolerable danger to the civilized world? If access to oil were of concern to them, one might have expected members of their administrations to hint as much. The [Margaret] Thatcher and [George H.W.] Bush administrations, after all, were quite open about the role that oil played in justifying the first go-around in Kuwait. Polls in the United States revealed at the time, moreover, that the public responded favorably to the argument. Why the supposed reticence now?

Regardless, it's difficult to know exactly what is being alleged when one is confronted by the slogan "No Blood for Oil!"

If the argument is that war is primarily being executed to ensure global access to Iraqi oil reserves, then it flounders upon misunderstanding. The only thing preventing Iraqi oil from entering the world market in force is the partial U.N. embargo on Iraqi exports. Surely if access to Iraqi oil were the issue, it would have occurred to Bush and Blair that removing the embargo is about $100 billion cheaper (and less risky politically) than going to war.

If the argument is that war is being undertaken to rape Iraqi reserves, flood the market with oil, bust the OPEC [Organization of Petroleum Exporting Countries] cartel, and provide cheap energy to western consumers, then war would be a dagger pointed at the heart of the "Big Oil" [multinational oil companies]. That's because low prices = low profits. Moreover, it would wipe out "Little Oil"—the small-time producers in Texas, Oklahoma, and the American Southwest that President Bush has long considered his best political friends. Accordingly, it's impossible to square this story with the allegation that President Bush is a puppet of the oil industry.

In fact, if oil company "fat cats" were calling the shots—as is often alleged by the protesters —President Bush would almost certainly not go to war. He would instead embrace the Franco-German-Russian plan of muscular but indefinite inspections because keeping the world on the precipice of uncertainty regarding conflict is the best guarantee that oil prices (and thus, oil profits) will remain at current levels.

If the argument is that "Big Oil" is less interested in high prices than it is with outright ownership of the Iraqi reserves, then how to account for Secretary of State Colin Powell's repeated promise that the oil reserves will be transferred to the Iraqi government after a new leadership is established? Do the protestors think that this high-profile public commitment is a bald-faced lie? Moreover, if that's the real goal of this war, then I'm forced to wonder why the U.S. didn't seize the Kuwaiti fields more than 10 years ago [during Operation Desert Storm].

Overstated oil profiteering

If the argument is that this war is aimed at installing a pro-American regime more inclined to grant oil contracts to American and British rather than French and Russian oil firms, then it invites a similar charge that France and Russia are against war primarily to protect their cozy economic relationships with the existing Iraqi regime. Regardless, only one or two American or British firms in this scenario would "win" economically while the rest would lose because increased production would lower global oil prices and thus profits. Because no one knows who would win the post-war contract "lottery," it makes little sense for the oil industry (or the politicians who supposedly cater to them) to support war.

The argument . . . that war is primarily being executed to ensure global access to Iraqi oil . . . flounders upon misunderstanding.

Moreover, the profit opportunities afforded by Iraqi development contracts are overstated. The post-war Iraqi regime would certainly ensure that most of the profits from development were captured by the new government, whose reconstruction needs will prove monumental. In fact, Secretary Powell has repeatedly hinted that Iraqi oil revenues would be used for exactly that purpose. Big money in the oil industry goes to those who own their reserves or who secure favorable development contracts, not to those who are forced to surrender most of the rents through negotiation.

If the argument is that the United States is going to war to tame OPEC (accomplished, presumably, by ensuring that a puppet regime holds the second largest reserves within the cartel), then it runs up against the fact that the United States has never had much complaint with OPEC. Occasional posturing notwithstanding, both have the same goal: stable prices between $20–$28 a barrel. The cartel wants to keep prices in that range because it maximizes their profits. The United States wants to keep prices in that range because it ensures the continued existence of the oil industry in the United States (which would completely disappear absent OPEC production constraints) without doing too much damage to the American economy. The United States doesn't need a client state within the cartel, particularly when the cost of procuring such a state will reach into the hundreds of billions of dollars.

Oil, however, is relevant to this extent: Whoever controls those reserves sits atop a large source of potential revenue which, in the hands of a rogue state, could bankroll a sizeable and dangerous military arsenal. That's why the United States and Great Britain care more about containing the ambitions of [Iraqi leader] Saddam Hussein. . . . Still, if seizing oil fields from anti-western regimes is the name of the game, why aren't U.S. troops massing on the Venezuelan border and menacing Castro "Mini-Me" Hugo Chavez?

In sum, the argument that the impending war with Iraq is fundamentally about oil doesn't add up. While everyone loves a nice tidy political morality play, I doubt there is one to be found here.

6

U.S. Dependence on Middle East Oil Threatens National Security

Conrad Burns

Conrad Burns is a U.S. senator from Montana.

As evidenced by the September 11, 2001, terrorist attacks against the United States, U.S. dependence on oil imports from rogue, undemocratic regimes in the Middle East—such as Saudi Arabia and Iraq—poses a grave national security threat. Oil profits can be diverted to radicals such as those responsible for the September 11 attacks. To lessen the influence that Middle Eastern nations have over the U.S. economy and foreign policy, the United States should buy more oil from Russia and the nations bordering the Caspian Sea. Most important, domestic sources of oil should be fully explored and developed, including the untapped oil and gas fields in Alaska, Montana, and other western states.

On September 11, 2001, we witnessed a catastrophic attack on our country, one that changed our outlook of the world in many ways. Americans realized for the first time that we did not live in a country that could escape the terrorist attacks that take place around the world.

The impact of oil dependency

We were not, and are still not living in a place immune to the terror some countries see on a daily basis. Despite the new realities we have seen since that fateful day, I have confidence in our government and know that since 9/11 many improvements have been made, and continue to be made, to make our country more secure.

In spite of these improvements, there is one area on which our country has neglected to take a strong stand: energy security.

Many have not realized the incredibly big impact that our oil dependency has on the security of our country. The attack on 9/11 by Islamic

Conrad Burns, "Beyond the Middle East: In Search of Energy Security," *Heritage Lectures*, March 19, 2003.

extremists should have been a wake-up call to the nation that our vital security interests are threatened by our increasing dependence on Middle East oil imports. I am sorry to say that our nation still slumbers. We should see the danger that lies in buying up to a quarter of our imported oil from Saudi Arabia and Iraq.

We should see the dangers of paying billions of dollars to a man [Iraqi leader Saddam Hussein] committed to amassing weapons of mass destruction.

We should see and understand that every time America buys a barrel of rogue oil we are in part funding unseen radicals.

And we should see that our national security is at risk, our foreign policy is shackled, and our diplomatic credibility in the Middle East undermined, so long as we buy from regimes that deny democracy and freedom.

America should not allow these regimes to maintain such a strong influence over our economy.

Our nation currently relies on foreign oil for 55 percent of our energy requirements, and this dependence is expected to rise to 65 percent by 2020. Indeed, during the next 20 years, our energy demand is expected to increase more than 50 percent. The United States uses oil to supply about 40 percent of our energy needs. No one could question the fact that energy is absolutely indispensable to maintain our national security and our way of life.

America has become increasingly dependent on petroleum from Saudi Arabia, Iran, and Iraq. With about 250 billion barrels, Saudi Arabia has more proven oil reserves than any other country; that is, one-fourth of the world's oil reserves.

Iran and Iraq each control about 10 percent of the world's oil reserves, with large amounts of unexplored resource. The expectation is that the Persian Gulf must expand oil production by almost 80 percent during the next two decades to supply the world market. That region has the natural resources and technical capability to achieve that production, but what will it cost us?

Our country is already too dependent upon foreign oil imports from the Middle East and this dependence is getting worse, not better.

[British prime minister from 1940 to 1945 and 1951 to 1955] Winston Churchill noted:

> On no one quality, on no one process, on no one country,
> on no one route, and on no one field must we be dependent.
> Safety and certainty in oil lie in variety and variety alone.

Diversity in supply and restraint in demand are the twin paths out of the crisis we face. Supply diversity can be achieved in two ways: by developing new international sources and by increasing domestic production.

Developing international sources and increasing domestic production

Internationally, our nation is ignoring the opportunities that lie in the Caspian Basin and Russia. In the Caspian Sea area, oil reserves of up to 33 billion barrels have been found, a potential greater than U.S. reserves and double those of the North Sea. Estimates are made of another 233 billion barrels in Caspian reserves. These could add up to 25 percent of the world's proven reserves. Russia may have even higher reserves.

Today, America buys virtually no oil from either the Caspian states or Russia. Massive investments are required to bring these resources on line, which means reforms will be needed in the nations bordering the Caspian Sea and in Russia itself. Private capital investment requires political stability and the rule of law. Contracts must be honored, corruption must disappear, and the regulatory regime must be favorable to attract investors. The governments in the countries concerned must have the political will to make the changes necessary to attract investment. But American leaders can, and ought to, encourage reforms more vigorously than we have.

Every time America buys a barrel of rogue oil we are in part funding unseen radicals.

Domestically, I believe it is important to find out what oil and gas resources we have access to in our own country.

Consider the contributions of my home state of Montana. Fuel cell research is underway at our universities. This will yield new ways to power our homes and cars. New technologies can turn Montana's abundant agricultural crops into alternative fuels.

Montana possesses the nation's largest supply of clean coal. We possess vast reserves of petroleum and natural gas.

If Montana has so much potential, think about our potential as a nation.

We must know where our reserves are, and we should have access to those resources to strengthen our own economy and energy security.

Access to federal lands should be a major goal of the domestic agenda. First on the list is opening the Arctic National Wildlife Refuge (ANWR), which would be a major improvement in energy position. This is not only an Alaska issue, it's an energy security issue.

Additionally, we must not overlook the great possibilities on federal land. There is a misconception that federal resources are being rapidly developed without regard to the land. The truth is that drilling activity continues to decline. In 2003, there are 30 percent fewer active drilling rigs than there were in 2001. The number in 2001 was 80 percent less than the drilling activity 20 years before. And currently, the federal government is actively producing natural gas from only 5 percent of its mineral estate. It is estimated that 95 percent of undiscovered oil and 40 percent of undiscovered gas is located under the lands in the Inter-Mountain West.

Environmental risks

Before 9/11, the U.S. Congress had the luxury of passing environmental laws that locked up huge areas on federal land and offshore. We could afford, some said, not to know what was "out there." The prevailing fear was that if we found oil or gas we might want to develop it. As it stands, less than 19 percent of the lower-48 [U.S. states] Outer Continental Shelf (OCS) lands [areas with offshore oil drilling potential] are available for development. This policy has led to absurd results. For example, our citizens in the Northeast are paying top dollar for Canadian natural gas taken from waters just to their north, while a moratorium exists on looking for the natural gas in similar continental shelf structures on the U.S. east coast. The National Academy of Sciences concluded that improved production technology and safety training of personnel have significantly reduced both blowout and operational spills on the OCS. The moratorium on OCS leasing is thus based on outdated facts and policy. I believe the environmental risks of transporting gas from the Middle East to our country are greater than bringing our OCS gas by submerged pipeline directly to shore

---

Our country is already too dependent upon . . . oil imports from the Middle East and this dependence is getting worse.

It also becomes quite clear that the economic and political benefits of using our own OCS gas resources are huge, and we must work to begin to take advantage of these possibilities. . . .

Energy security is a complicated topic and I fully realize that my brief [analysis] only touches on a few aspects of the problem. I hope that you understand my conviction that we can no longer continue with business as usual on foreign oil imports. We must all realize the dangers that lie in our current energy dependencies, and we must work to make changes to create energy security for our country.

Since 9/11 we have come to live in a world with new threats and new opportunities. We have come to live in a world where a new reality has shown us that energy diversity is not simply a good idea, but it is essential to our national survival.

7

U.S. Dependence on Middle East Oil Is Not a National Security Issue

Donald Losman

Donald Losman is a professor of economics at the Industrial College of the Armed Forces, a school within the National Defense University located in Washington, D.C.

Since the mid-1970s, the United States has used military force to secure unfettered access to imported oil. Proponents of current U.S. foreign policy argue that stable access to oil is a matter of national security. They maintain that the oil embargo of the early 1970s, imposed by oil-producing states in the Middle East, caused a severe economic crisis in the United States that threatened national security. The economic downturn of the early 1970s, however, was essentially caused by the misguided regulatory policies of the U.S. government, which caused oil prices to artificially inflate. Not only is the chance of a significant denial of oil to the United States very low, but addressing oil-supply disruptions through military means is costly, immoral, and inappropriate. Effective market solutions, rather than the spilling of American or foreign blood, will ensure that oil prices remain stable.

E conomic security issues have traditionally centered on health uncertainties, retirement needs, and protection against income interruption. However, over the past quarter-century, the U.S.'s civilian leadership and the military community, joined by a variety of domestic groups, have transformed the concept of economic security into a prominent national security issue. Undoubtedly, the major impetus for this was the 1973 OPEC [Organization of Petroleum Exporting Countries] oil embargo and the economic trauma of the 1970s. In those dark days of oil shortages, record interest rates, and rapacious inflation, foreign economic "weapons" appeared to threaten the economic well-being of the U.S. and possibly even jeopardize important strategic interests. In debates on national security, economic concepts and references began to abound. Reflecting this blending of economics and national security, public opinion pollsters began to ask Americans what their perceptions were of the gravest national security threats, usually posing economic challenges (such as those from Japan) as one choice and military threats (from the Soviet Union or North Korea, for example) as another.

Military force for economic goals

Nevertheless, even as late as 1987 and 1988, formal documents on national security strategy remained narrowly focused on military power and the U.S. rivalry with the Soviet Union. The documents began to broaden, though, particularly in the administrations of George H. Bush and Bill Clinton, which emphasized the role of economics and entertained the inclusion of environmental policy. For instance, the first page of the introduction to *A National Security Strategy for a New Century*, published in December, 1999, has the word "economy" or "economic" five times, and "prosperity" appears twice. Listed under vital national interests is the "economic well-being of our society." The paragraph concludes, "We will do what we must to defend these interests . . . using our military might unilaterally and decisively," if necessary.

The link between imprudent American economic policies and the subsequent [1970s oil] crisis cannot be overstated.

Those statements indicate that America is willing to use military force to attain economic goals. Contrast that text with the 1988 report's introduction, in which the word "prosperity" never appears, and "economic" is used just three times, mainly as a tool to achieve larger ends, rather than as an end in itself.

Rarely, however, has the economic content of national security policy been put to a rigorous intellectual or logical test. Instead, it has simply been accepted. Yet, the economic security concept as a national security goal is ill-suited, imprecise, and unnecessarily costly, and could entail using U.S. military might in dubious ventures. Moreover, attaining economic objectives through the use (or threatened use) of military force is essentially a "might makes right" philosophy. At best, it is morally questionable; at worst, abhorrent. Operationally, the concept causes problems that complicate and degrade appropriate national security missions.

Human beings tend to be averse to risk, and economic insecurity has traditionally been addressed by economic measures. It has also been addressed, rightly or wrongly, in government programs such as Social Security, Medicare, unemployment compensation, welfare, and trade protection. In short, the U.S. has used private or public economic measures, not the military, as the main provider of economic security.

A specter is haunting America—the continuing and, at times, almost hysterical fear of oil shocks. In October, 1973, the Arab oil-producing states imposed production restraints and an embargo—their second such attempt. They did so allegedly as a punishment for those countries that supported Israel in the Middle East war earlier that month. Their first effort at embargo occurred in 1967, following the lightning Israeli victory in the June Six-Day War.[1] That episode is not well-known because it was a total failure. However, global oil market conditions would change substantially in subsequent years.

Oil in the U.S. had been governed by a maze of state and Federal regulations. As a subsidy to domestic oil producers, nominal American prices were held relatively stable—and higher than world prices—from the 1950s through 1973. It was not until 1974, after the second OPEC embargo, that inflation-adjusted (real) U.S. prices were kept below world levels. Nonetheless, even before 1974, domestic price signals were misleading and promoted vulnerability to price and supply disruptions. For example, between 1970 and 1971, inflation-adjusted domestic crude oil prices declined 1.2%, despite real world prices rising more than 21.2%. From 1971 to 1972, real U.S. prices again declined (3.4%) in the face of another increase (7.9%) on world markets, thereby giving American consumers and businesses the illusion of greater availability of oil when just the opposite was occurring.

Internationally, excess producing capacity outside the OPEC countries had virtually disappeared by 1970, just as production had peaked in the U.S. and Canada. Libya's successful negotiations in 1970 with major oil companies marked the beginning of a significant shift in power between the international oil companies and the Middle East oil-producing states, with the latter enhancing their bargaining positions substantially.

Accordingly, economic development around the world, coupled with U.S. government-manipulated domestic oil pricing, brought a serious vulnerability to energy-importing nations in general and to America in particular. The U.S. built a society that resided far from work, drove gas-guzzling automobiles, and lived in energy-inefficient homes. The oil shock of 1973 was extremely disruptive and raised energy and related prices in an American economy that had already been steadily inflating since the mid 1960s. In August, 1971, more than two years before the oil shock, President Richard M. Nixon had invoked price controls to contain inflation. Although the Netherlands and the U.S. were the two targets of the embargo, the entire world was hit with sharply higher oil costs as a result of the 1973 oil crisis.

Fear [of Middle East oil shocks] has fueled questionable changes in national security perspective.

The link between imprudent American economic policies and the subsequent political and economic crisis cannot be overstated. Indeed, an "energy crisis" was developing long before the October, 1973, war. In 1971, in Teheran [capital of Iran], the OPEC producers negotiated a five-year agreement with the oil companies for higher prices, but, in the face

---

1. In that war Israel defeated four neighboring Arab states and gained territory including the West Bank, Golan Heights, and the Gaza Strip.

of swollen demand, it was almost immediately abrogated.

Although numerous reasons for the tight market for oil existed, the most important, by far, was the rapidly growing U.S. demand. By 1972, U.S. consumption was already pushing the world demand beyond planned production, refining, and transport capacities. By late spring, 1973, oil company executives were warning of a coming crisis, but the American government seemed to discount the seriousness of the threat. By summer, OPEC was calling for a conference to amend the five-year agreement again.

America's military might can do little to bolster structurally twisted and institutionally flawed economies [of oil-producing countries].

When war broke out in the Middle East in October [1973], an OPEC delegation had been in Vienna [Austria], negotiating with the big oil companies for yet another round of price increases. The timing was fortuitous for Arab suppliers, who would drape their price-raising production cutbacks in the flag of Pan-Arab rhetoric. Acting opportunistically, non-Arab cartel members went along for the ride. The oil companies were unable to respond, while their governments' prime concern focused on "cooling" the Arab-Israeli conflict and avoiding a superpower confrontation. So, the October price increases were followed by another set of hikes in December. In a very short time, world oil prices had jumped enormously. The price refiners paid for crude oil, for instance, averaged $3.58 a barrel in 1972. By 1974, the corresponding figure was $9.07.

Instead of abandoning the price control system, which had led to U.S. vulnerability, the government merely raised price ceilings and resorted to a variety of rationing devices and mandatory allocation schemes to manage the shortages created by continuing controls. Paul MacAvoy, author of *Energy Policy: An Economic Analysis*, summarized the U.S. economic policy response:

> While the world prices went up more than 200 percent, United States crude prices increased only 56 percent because of federal ceilings on domestic product prices. . . . The controlled domestic prices increased slowly. . . . United States crude prices never caught up with, but rather were at 60 percent of, the international crude prices at the end of the decade [1970s].

Despite the embargo, U.S. oil stockpiles fell only slightly, and, by March, 1974, they were growing again. At the time, that fact was not widely known by the public, nor was it a source of great comfort because future national needs, prospective price increases, and other variables were totally unknown. Indeed, the OPEC embargo was not terribly effective since supplies meant for other consuming countries were diverted to the U.S. However, bureaucratic errors worsened America's situation. As MacAvoy noted, "regulation created the effects of the embargo . . . and the FEO [Federal Energy Office] gets the credit for the energy crisis perceived by consumers in 1974."

The U.S. government attempted to counter the negative effects of the oil price shock with expansionary monetary and fiscal policies designed to reduce—through high levels of aggregate spending—the unemployment associated with the oil price shock. A variety of jawboning public affairs campaigns cajoling business and labor to hold prices and costs down were employed as well, all to little effect.

Major oil price increases again occurred toward the end of the 1970s. These had several sources. Domestically, demand remained artificially stimulated by continuing price controls. Their partial easing in 1979, though, provided a short-term impetus to higher domestic prices. Second, having witnessed real oil prices eroded by continuing global inflation, OPEC attempted to implement production cutbacks. Such price-increasing endeavors were greatly assisted by the Iranian revolution, which probably disrupted world oil supplies twice as much as did that associated with the October, 1973, Arab-Israeli War.

Finally, all of those actions fueled a speculative panic in global oil markets, which itself became yet another source of higher prices. Accordingly, severe recession and rapacious inflation—and their mental association with oil price hikes—seem to have instilled a permanent and lurking fear in American policymakers and the public. That fear has fueled questionable changes in national security perspective.

Government policies to blame

Far too much of the economic debacle of the 1970s has been attributed to OPEC and the price hikes and far too little to U.S. government policies—both pre- and post-embargo. Economist Douglas Bohi estimated that the petroleum shortages of the 1970s reduced gross domestic product just 0.35%. At the macroeconomic level, the combination of easy money and budget deficits virtually guaranteed that inflation, high interest rates, and financial dislocations would worsen—thus adding to the damage that OPEC originally imposed. Indeed, economist John Kenneth Galbraith correctly pointed out that the recipients of swollen OPEC revenues—by pulling enormous amounts of purchasing power out of the U.S. and putting less back into the American economy through their purchases—had actually imposed a deflationary pressure. Without government efforts, some degree of deflation—not inflation—would have resulted.

> *The best way to address [the] economic challenges [of dependence on Middle East oil] is through good economic policy, not military means.*

According to Jack Guynn, president of the Federal Reserve Bank of Atlanta, "After the major oil price shocks . . . the Fed eased monetary policy . . . and the inflation that ensued was entirely predictable. But it was the monetary policy response, not the oil price increase, that led to inflation." Eugene Guccione, an energy specialist, succinctly noted that the energy crisis was triggered by the government's good intention, the idea that energy should be cheap, and worsened by its use of compulsion—

government price controls—to achieve that goal.

Even today, these facts are hardly common knowledge. Thus, the catastrophic economy of the 1970s and early 1980s is still attributed almost entirely to oil shocks. In the post-1973 period, the specter of nefarious Middle Eastern sheiks again clubbing the U.S. with their oil weapon or extracting diplomatic and foreign policy concessions merely by implying that threat seemed to become embedded in America's national psyche. Additionally, the 1970s ended with further significant oil price increases as the U.S. lost Iran—one of its "twin pillars" [along with Saudi Arabia] in the oil-producing region—to forces seemingly opposed in every way to American interests and values. The 1980s then began with a severe—albeit brief—recession, which was shortly followed by the most significant economic downturn since the 1930s. Finally, the nation was regularly exposed to a large and vocal body of alleged "experts" whose calamitous lamentations alternated between "the world is running out of oil" and "OPEC is going to get us again." As a result, the fear became embedded, and oil, a disastrous economy, and national security became intertwined.

Fear and fiction surround Middle East oil supplies

In what was widely interpreted as a saber-rattling exercise, one later repeated by Secretary of State Henry Kissinger [who served under presidents Richard M. Nixon and Gerald R. Ford], President Gerald R. Ford declared in September, 1974, "Throughout history, nations have gone to war over natural advantages such as water or food." America was uneasy, and politicians sensed that the public wanted its fears addressed. By the mid 1980s, the world economy was increasingly interconnected, and some people were afraid of U.S "dependence" on goods from other countries. It did not take policy analysts—and others—long to embellish further the economic security argument with suggestions that faltering overseas economies were ripe for communism or radical Islamic fundamentalism or that glaring global income inequalities might easily plunge have-not nations into war against the "haves."

Oil disruptions can be mitigated by . . . stockpiling, futures contracts, diversifying the supplier base, and relaxing regulatory restrictions.

Failing overseas economies may indeed become politically unstable and susceptible to radicalism, either secular or religious, but America's military might can do little to bolster structurally twisted and institutionally flawed economies. On the contrary, economic problems, both here and abroad, cry out for economic solutions—not military ones. The kernel of truth—in both the "faltering economies breed radicalism" argument and the U.S. economic vulnerability lamentation—has been mixed with a great deal of fear and fiction.

Thus inspired, the U.S. military fashioned, in the late 1970s, a Rapid Deployment Joint Task Force, the precursor to the Central Command (CENTCOM), to "protect" Middle Eastern oil supplies and their somewhat precarious oil-producing regimes. Indeed, by the second half of the 1980s, the merger of economics and national security in the public's mind was so far advanced that several opinion polls found many U.S. citizens believing that Japan was more of a threat to America's security than was the Soviet Union.

Finally, many independent organizations involved in national security issues as well as think tanks specializing in the Middle East reflected and abetted this merger of economic security and national security. Many of the think tanks were traditionally regarded as somewhat obscure scholarly organizations often deemed to be studying arcane civilizations in a deserted corner of the globe. Suddenly, oil became universally acclaimed as a strategic good. Television interviews were requested of scholars of the Middle East, and microphones were shoved in their faces. Oil and perceived oil power had given their corner of the world the importance that they believed it truly deserved. Their region was transformed from relative obscurity to a most-visible, vital U.S. interest.

A call for economic solutions

As is so often the case, actions and policies inspired by fear were terribly misguided. As noted, most of the economic ills of the 1970s and early 1980s derived not from oil shocks per se, but from inappropriate monetary and fiscal policies and other policy failures. The economic costs of the oil shocks were nowhere near as great as imagined, and the world has not run out of oil or any other critical commodity. Indeed, virtually all commodities are more abundant today, as attested to by their lower prices, than they were 30 years ago, thanks to advancing technologies and market-directed allocations.

The best way to address economic challenges is through good economic policy, not military means. Stockpiling, more realistic prices for oil, incentives for energy conservation that such pricing provides, and similar measures improve supply and demand outcomes and reduce national security vulnerability. Years after the oil embargo, a variety of useful economic and energy policy changes—including price deregulation—created a new environment for oil pricing. Today, nominal oil prices, seemingly so (and, to some, unacceptably) high, are about the same level as in 1990–91 during the Kuwait crisis,[2] and lower than in the 1985–86 period.

Inflation-adjusted numbers are even more revealing. With just a few exceptions, the constant dollar cost of energy has declined almost continually since 1981. For oil in particular, the average inflation-adjusted U.S. price in 1999 was lower than all but two years since 1974, and the relatively depressed 1998 cost was the lowest real price in more than 50 years. Moreover, the amount of oil consumed per billion dollars in real gross domestic product declined nearly 37% between 1973 and 1993, a trend that has continued and is likely to accelerate as the Internet economy spreads. Finally, when measured in terms of work time at the average manufacturing wage, the price of gasoline has been declining almost continually since the 1920s. In 2000, even with relatively recent price increases, the "aver-

---

2. reference to the 1991 Persian Gulf War precipitated by Iraq's invasion of Kuwait

age work time needed to buy a gallon of gasoline rose just two minutes from its all-time low of 4.2 minutes at the end of 1998," as W. Michael Cox pointed out in *Investor's Business Daily*, November 20, 2000.

Although neither Iraqi dictator Saddam Hussein [overthrown in April 2003 as a result of the U.S.-Britain invasion of Iraq] nor Iran's rulers find favor with the U.S., they are desperate to sell their oil. Indeed, in 1998, Iraq was America's eighth-largest supplier of crude oil. The entire world "drinks" from what is essentially one global pool: any refusal to sell to the U.S. while supplying other countries simply releases some other seller's oil to be sold to America. In any case, the U.S. is much less directly dependent on Middle East supplies today than it was in the early 1970s. Venezuela and Saudi Arabia tend to alternate as our number-one and -two suppliers, with Mexico in third place and Canada just behind it.

> *It is one thing to fight for . . . human rights . . . ,*
> *but spilling blood to ensure that people get cheap*
> *[oil] . . . is quite another story.*

In short, today's realities are light-years away from the situation in the 1970s and the chilling future scenarios that were depicted in those dreary days. Indeed, the importance of oil to economic growth has undergone recent "laboratory" testing. The price hikes of 1999–2000, coupled with a serious decline in the foreign exchange value of the euro [the currency of the European Union], imposed a crude oil price increase on Germany of 211% from the fourth quarter of 1998 through the third quarter of 2000. Nonetheless, economic growth—with falling inflation and a drop in unemployment—proceeded. Yet, America steadfastly clings to perceptions formed in the 1970s, and national policies continue to reflect oil paranoia. This inertia seems like another example of preparing for the last war.

Mitigating oil disruptions

Even more important, there are standard economic ways and means to address shortages and reliability of supply issues. Using or threatening to use military force is costly, cumbersome, and inappropriate. If supplies of vital items tighten and their prices rise, the U.S. economy will, of course, be adversely affected and forced to make adjustments to mitigate losses. When such discomfort emanates from crop failures, labor problems, bad weather, earthquakes, etc., Americans do not normally feel compelled to call on the national security establishment to address the situation. Society lets the economy address such issues, in full recognition that to do otherwise would probably degrade other military missions and achieve the economic goals very inefficiently, if at all.

Oil disruptions can be mitigated by such means as stockpiling, futures contracts, diversifying the supplier base, and relaxing regulatory restrictions. Allowing higher market prices will automatically produce increased supplies and simultaneously encourage judicious usage. The "pain" of such price hikes can be substantially soothed by reduced gasoline taxes—

the average combined Federal and state tax per gallon of 38 cents allows plenty of "wiggle room."

Effective market solutions—rather than military ones—are readily available. In a market system, they are used to prevent shortages of goods and commodities. Why, then, is oil treated differently? Clearly, the specter of the 1970s is still haunting America. It must be exorcised.

U.S. policymakers need to ask whether it is worth spilling American (or foreign) blood to keep commodity prices at "acceptable" levels. If the U.S. is willing to risk the lives of Americans to ensure adequate oil flows at reasonable prices, perhaps it should be willing to do so for access to coffee. U.S. society values coffee more highly than petroleum, if one compares their relative per-unit prices.

An argument can be made that coffee is a luxury, so that an interruption in its flow would not be that disruptive. Oil, however, is a critical input to the whole industrial establishment. Although that is true, semiconductors would also clearly fall into the critical input category—as would a number of raw materials. Indeed, America's expenditures on semiconductors are approaching $200,000,000,000 annually, about 35% more than the U.S. spends on oil. Semiconductors and other electronics components are vital to all of the economic sectors in the new information economy. Even the automotive industry, an "old economy" industrial sector, consumes more than five percent of the semiconductor output.

Given the crucial role of semiconductors in the economy, as well as their importance for national security (the Pentagon has numerous systems dependent on them), it would not be difficult to conjure up a worst-case scenario similar to that involving oil. After all, the U.S. imports significant amounts of semiconductors, and about 80% come from the Far East. Around 72% of computer mouses and 65% of keyboards come from just one Asian nation (Taiwan). Moreover, that region has regimes hostile to U.S. interests and is subject to various kinds of instability.

If the U.S. is willing to put its people in harm's way for oil, why not for electronic and other "critical" imports? A very large number of goods contribute significantly to the "economic well-being of our society." America cannot use military power to ensure access to all of them. It is no easy job to discern which commodities are the most significant. Defining economic goods as "strategic" is always difficult and becomes even more so as technology progresses and the world becomes more complex and interdependent.

Questioning the spilling of blood for oil

Perhaps even more important than earlier arguments are the moral considerations surrounding whether the U.S. should attempt to guarantee access to resources by using military means. The moral issues are nothing less than appalling. For instance, in July, 2000, in a lecture to about 300 military reserve officers, I mentioned high gasoline prices in the Midwest and asked how many from that region were just "dying to have lower gas prices." Many hands were raised. I then rephrased the question, asking. "How many of you would be willing to die for lower gas prices, or put your children's lives at risk?" Although the audience was a very patriotic group, not a single hand was raised. Indeed, the rephrasing casts the issue in a completely different and more appropriate light—blood for oil. If viewed from this more sobering perspective, the public's cost/benefit calculations are likely to change.

Higher oil, coffee, or semiconductor prices would reduce the real standard of living in the U.S. America's economic well-being would be slightly impaired. Nevertheless, putting members of the armed forces in harm's way in order to protect Americans' standard of living, much less merely one component of that standard of living, is questionable. It is one thing to fight for freedom and human rights or to deter and defeat military aggression, but spilling blood to ensure that people get cheap goods—for instance, to ensure their right to Sunday driving—is quite another story. Many, nonmilitary options exist for mitigating the pain and supply disruptions, including allowing the market to reallocate resources. It is wrong to use U.S. armed forces against people in foreign nations so that Americans can get a better deal.

Such a concept is morally objectionable, as demonstrated by plenty of historical precedents. For example, Imperial Japan's [1868–1912] conquests under the euphemistic "coprosperity sphere" were driven mainly by fear of losing raw material supplies for its industrial establishment. Such motivations and rationalizations, so morally unacceptable to Americans, are today accepted as part of U.S. national security strategy. Indeed, in the 1997 document, the "free flow of oil" was not only deemed essential, it had to be at "reasonable prices." Yet, few people in the U.S. have been outraged, and there has been no groundswell of moral indignation.

The likelihood of a significant denial of oil [by Middle Eastern states] to the U.S. is very low.

Venezuela—one of the U.S.'s top-two overseas oil suppliers—is led by an unpredictable, left-leaning government whose leader, President Hugo Rafael Chavez Frias, has consorted with the likes of [Iraqi leader until 2003] Saddam Hussein and Cuba's Fidel Castro [Communist leader hostile to the United States]. A scenario involving major oil price hikes by Venezuela is not unimaginable. Since "unreasonable" prices would probably emerge from the scenario, the 1997 national security strategy shows a willingness to send in troops. Such an action would not be in consonance with American—or global—norms of morality. Neither would it enhance the diplomatic prestige or the global leadership role of the U.S. It is therefore natural to wonder why defending oil is in America's national security strategy.

A wasted effort to safeguard Middle East oil supplies

Vital national interests are difficult to define, which leads to continuing efforts to expand their scope. For instance, in 2000, Vice President Al Gore and environmental groups pushed to have the cessation of irreversible, large-scale environmental damage listed as a vital national interest. The Joint Chiefs of Staff [U.S. military body composed of the secretary of defense and the National Security Council] opposed that expanded interpretation. More recently, efforts were made by African regional specialists to have access to important mineral resources declared a vital U.S. interest. The Department of Defense and economists from the Clinton Administration rejected that attempt.

Every extra theater of operations and every new mission complicate the military planning process and draw resources away from competing demands. An estimated $30–60,000,000,000 a year has been expended to safeguard Middle East oil supplies, even as oil has become increasingly abundant (as signaled by the long-term decline in prices). That cost is high absolutely and is enormously high relative to the total value of U.S. oil imports from the Persian Gulf, which averaged $10,250,000,000 annually in the period from 1992 to 1999.

The financial commitment is a huge burden for the Pentagon [headquarters of the U.S. Department of Defense]. Tradeoffs between the readiness, sustainability, and modernization of the armed forces have become much more acute because resources that might have been allocated to these ends went elsewhere. Operation tempos have substantially increased throughout the services as they have stretched their resources—thus harming morale, operational capabilities, and readiness, as well as exacerbating retention problems.

Strategic direction becomes muddled when national security, prosperity, human rights, the role of law, and environmental concerns are wrapped into one package. U.S. national security documents do not list priorities—thus implying an equality of interests, which is surely not the case. Moreover, "bolstering" or "maintaining" prosperity is a broad goal and might mean that anything that diminishes American affluence is a national security issue. Accordingly, if access to oil or minerals is really a vital national interest, the U.S. military must be able to respond to instability in each and every region of the world. That requirement is simply too ambitious, and even the attempt to meet it dilutes military resources and planning capabilities.

Formal published declarations of American resource needs (for instance, the "free flow of oil" as a vital national interest) are likely to increase the probability of the contingencies the U.S. is hoping to avoid. To the degree that the U.S. formally announces oil security as a vital national interest, it tacitly declares the vulnerability of its oil supply. Unfriendly elements might come to believe that an oil disruption anywhere will seriously damage either the American economy or U.S. national security posture—exactly the scenario Washington wishes to avoid.

If, on the other hand, economic approaches are undertaken to ensure resource supplies—rational pricing, diversified sourcing, and market-determined resource allocations—the nation will be less vulnerable. Defending resources such as oil would not need to be part of the national security strategy. Moreover, the incentive for foreign nations to threaten American oil supplies would be relatively low because the likely long-term damage to the U.S. would be limited.

Time for a reassessment

The likelihood of a significant denial of oil to the U.S. is very low, despite the conflict in Afghanistan [where the U.S. military has taken action against terrorists] and any possible future actions against Middle Eastern states abetting terrorism. Market forces are generally sufficient to prevent a serious crisis. There is no need to use military force or to divert large portions of Pentagon resources to guarantee cheap oil or to achieve any other economic end. As Barry Buzan, the author of *People, States, and Fear,* has noted. "Although the case for economic threats to be counted as threats to national security is superficially plausible, it must be treated with considerable caution." That caution has been lacking in recent decades.

Such a line of reasoning does not deny that history is replete with economically motivated military onslaughts, such as Saddam's invasion of Kuwait [in 1990]. The immorality of such actions is universally recognized. Yet, when a democratic superpower espouses strategic euphemisms to veil similar military interventions, the costs and the immorality become obscured. It is time for America to end its paranoia about fuel supplies. If the U.S. really wants economic security, there are economic ways to attain it. Economic solutions will cost far less—when all human, military, and financial costs are totaled—than military solutions.

Economics has an important role in grand strategy. In America's national security strategy, though, economics should factor in as a funding constraint, a tool for resource allocation, and a vehicle to assess the costs of America's goals, options, and alternative national security strategies.

The elimination of prosperity and economic security from the U.S.'s national security strategy "to do" list will eliminate some costly financial drains and remove significant "clutter" from an inherently difficult process. Real national security will be enhanced by eliminating unimportant goals and focusing on more important matters. America is in the 21st century, not the 1970s. It is time to reexamine and reformulate U.S. goals in national security policy.

8

Raising Car and Truck Mileage Standards Will Reduce Oil Imports

David Friedman

David Friedman is the senior analyst in the Clean Vehicles Program of the Union of Concerned Scientists, an organization involved in solving environmental problems using scientific and technological solutions.

Forty percent of the 10 million barrels of oil and petroleum products that the United States imports each day is consumed by motorists. Close to 25 percent of this oil comes from the politically unstable Middle East, a fact which compromises U.S. foreign policy and the government's efforts to defeat international terrorism. To reduce foreign oil dependence as well as the pollution emitted by cars and trucks, the federal government should raise the corporate average fuel economy (CAFE) standards—the mile-per-gallon standards with which car manufacturers must comply—to 40 miles per gallon (mpg) by 2012. This phased-in increase would save 125 billion gallons of gasoline by 2012, about 1.9 million barrels of oil per day.

U S drivers consumed 121 billion gallons of gasoline in 2000 at a total cost of $186 billion. This level of consumption represents 40 percent of the oil products that the nation consumes. This number places these vehicles at the heart of the growing debate over oil supplies.

The importance of fuel economy

Today, US oil dependence is greater than it has ever been as we import a record 10 million barrels of oil and petroleum products each day. These imports represent over half of US oil product consumption, and as demand increases the proportion of imports will rise. About 25% of this imported oil comes from the politically unstable Middle East—for example in the year 2000 we imported 1.7 million barrels of oil per day from Saudi

David Friedman, statement before the Committee on Commerce, Science, and Transportation, U.S. Senate, Washington, DC, November 6, 2001.

Arabia and another 0.6 million barrels per day from Iraq. The cost of imported oil exacts a toll on our international balance of trade, as the United States currently sends about $200,000 overseas each minute to buy oil products and is estimated to spend $20 to $40 billion per year to defend oil resources in the Middle East.

In recent years, the Organization of Petroleum Exporting Countries (OPEC) [cartel of oil-producing countries] has regained its ability to substantially influence the price of oil throughout the world. OPEC's market power can be expected to grow as its production approaches half of all world oil output in the next two decades. In the United States, our dependence on imported oil from OPEC and other foreign sources is expected to grow to 64%, making us even more susceptible to supply shortages and rapid rises in world oil prices. Historically oil price shocks and periods of inflation have coincided, resulting in significant harm to the US economy and our balance of trade.

[US] dependence on imported oil . . . is expected to grow to 64%, making [Americans] even more susceptible to supply shortages.

Transportation is also the source of roughly one-third of all the heat-trapping gases (greenhouse gases) linked to global warming that are released in the United States every year. Greenhouse-gas emissions from the US transportation sector amount to more than most countries release from all sources combined. The production, transportation, and use of gasoline for cars and light trucks resulted in the emission of 1,450 tons of greenhouse gases by the United States in 2000—over one-fifth of US global warming emissions that year.

Cars and trucks are the second largest single source of air pollution in the country, second only to electricity generation. As tailpipe standards are tightened, pollutants from passenger vehicles are falling to near the level of those produced in refining and distributing gasoline. As a result, transportation's impact on air pollution will soon approach an equal split between the tailpipe and the amount of fuel a vehicle uses. In the case of toxic emissions, pollutants that may be linked to cancer, the upstream emissions from fuel refining and distribution are the dominant source. The production and distribution of gasoline is also linked to many other negative environmental impacts including oil spills and groundwater pollution.

Assuming current fuel use, the production and distribution of gasoline alone results in the emission of 848,000 tons of smog-forming pollution and 392,000 tons of benzene-equivalent toxic emissions in the United States each year. Reducing these numbers significantly through improvements in fuel economy can mean great strides in protecting human health.

The pitfalls of stagnant fuel economy standards

The effect our cars and light trucks have on our economy, our oil use, and our environment is only expected to get worse due to rising vehicle travel, a changing vehicle fleet, the impacts of vehicle emissions and fuel use under actual driving conditions, and stagnant fuel economy standards. Together these factors have led to a 24 mpg fleet average fuel economy in 2000, the lowest level in over twenty years:

- *Rising Travel.* There are now more vehicles in the United States than people licensed to drive them. Combined with increasing travel rates per vehicle, the number of miles that Americans are driving continues to rise. Vehicle travel is expected to increase nearly 50% over the next 20 years, a trend that will help drive up passenger vehicle fuel use.
- *Shifting Markets.* SUVs [sport utility vehicles] and other light trucks are allowed to use one third more than cars under current CAFE requirements [miles-per-gallon fuel economy standards set by the federal government]. This "Light Truck Loophole" caused consumers to use about 20 billion more gallons of gasoline in 2000 and cost consumers about $30 billion dollars more than if the fuel economy standards of light trucks was set to the same as that of cars. The light truck market has risen from 19% to 46% since 1975 [the year CAFE was enacted] and is expected to grow to at least 50% of the passenger vehicle market, driving fuel economy lower in the coming years.
- *Real World Fuel Economy.* Testing for CAFE standards is based on a pair of simulated driving cycles established in 1975. At the time it was unclear if these cycles represented real world driving conditions, but today it is quite clear that they do not. Estimates show that real world fuel economy is about 17% below tested values and this shortfall is expected to increase over the next two decades.
- *Stagnant Fuel Economy Standards.* CAFE standards for cars and light trucks have not changed in more than a decade.[1] The original schedule called for an increase in car fuel economy to 27.5 mpg by 1985. While this goal was delayed for a few years, the standard has been at that level since 1990. The light truck standard reached approximately today's level in the late 1980s while separate standards existed for 2 and 4-wheel drive vehicles, and like passenger cars, was stalled for a short period until reaching today's 20.7 mpg requirement.

We estimate that these factors, along with continued stagnant fuel economy standards, would lead to an increase in passenger vehicle fuel use over the next two decades of 56 percent, to 189 billion gallons per year, by 2020. The result would be fuel costs to consumers of $260 billion dollars at a gasoline price of $1.40. Total oil demand would rise from [2001's] 20 million barrels per day to over 27 million barrels per day by 2020, 64% of which would be imported form outside the US. In addition, annual greenhouse-gas emissions from the passenger vehicle sector would rise to 2,260 million tons of carbon dioxide equivalent while emission of 1,320,000 tons of smog-forming pollutants and 612,000 tons of benzene-equivalent toxic emissions would be produced in the United States each year.

---

1. In June 2002, the U.S. Senate voted to remove a provision from energy bill S.517 that would have raised average car mileage requirements from 27.5 mpg to 36 mpg by 2015. In January 2003, the administration of President George W. Bush increased the government mandated standards for fuel efficiency in light trucks, a category that includes sport utility vehicles, from the current standard of 20.7 mpg to 22.2 mpg for model year 2007.

Reforming regulations to reduce the impacts of driving

The US is not locked into the predictions noted above. A systematic approach to reducing fuel use would address all of the key factors noted above: stagnant fuel economy standards, shifting markets, real world fuel economy, and rising travel. Within this systematic approach, increasing fuel economy standards to 40 mpg by 2012 is the single most effective, fastest and least expensive path to reducing our future dependence on oil.

The 2001 National Research Council [NRC] study has identified the CAFE standards enacted in 1975 as a key factor in the near doubling of new passenger car fuel economy (15.8 mpg in 1975 rising to a peak of 28.5 in 1998) and the 50% increase in the fuel economy of new light trucks (from 13.7 mpg in 1975 to today's 20.7 mpg). In addition, this study notes that CAFE standards have played a leading role in preventing fuel economy levels from dropping as fuel prices declined in the 1990s. UCS [the Union of Concerned Scientists] estimates that current fuel economy levels maintained by CAFE saved consumers over $90 billion in 2000. The NAS report estimates that in the year 2000 alone, increased fuel economy reduced gasoline use by 43 billion gallons, or about 2.8 million barrels of oil per day (UCS estimates the figure to be about 60 billion gallons of gasoline, or 3.9 million barrels of oil per day).

Stagnant fuel economy standards . . . have led to . . . the lowest [average fuel economy ratings] in over twenty years.

These savings put to rest concerns over the effectiveness of improved fuel economy. While fuel use has risen by 30% since the CAFE law was passed, this is primarily due to an increase in the amount of travel by Americans each year—which would have resulted in an even larger increase in fuel use had vehicle fuel economy not improved.

Savings of the same magnitude as seen in the past can be achieved in the future if fuel economy standards are again increased. UCS analysis has shown that cost-effective technologies for near-term and longer-term improvements in vehicle efficiency exist today. If these technologies are used to increase fuel economy over the next 20 years, our passenger vehicle oil use could be turned around (i.e. we could stop the growth in fuel use and even turn back the clock to 1990 levels if standards are raised sufficiently), the amount of money consumers spend on gasoline could be substantially reduced, and the impact our driving has on the environment could be cut in half. Table 1 is a short list of conventional technologies that have already been developed by automakers that could significantly increase the fuel economy of today's cars and light trucks, many of which are already in some cars today.

Estimates from a [2001] study released by the American Council for an Energy Efficient Economy [ACEEE], by John DeCicco, Feng An, and Marc Ross indicate that a combination of these technologies, along with mass reductions targeted at the heaviest vehicles, can produce a fleet of cars and trucks that averages over 40 miles per gallon. Table 2 shows the costs and

Table 1: Existing Conventional Technology Options for Fuel Economy Improvement

Vehicle Load Reduction
- Aerodynamic Improvements
- Rolling Resistance Improvements
- Safety Enhancing Mass Reduction
- Accessory Load Reduction

Efficient Engines
- Variable Valve Control Engines
- Stoichiometric Burn Gasoline Direct Injection Engines

Integrated Starter Generators

Improved Transmissions
- 5- and 6-speed automatic transmissions
- 5-speed motorized gear shift transmissions
- Optimized shift schedules
- Continuously Variable Transmissions

net savings associated with these improvements in fuel economy. The result is an increase in fuel economy of over 70% and a net saving to the average consumer of over $2,000. Increasing fuel economy standards results in a win-win situation where consumers and the environment are both better off. In this case, fuel economy standards result in a net cost of car-

Table 2: Fuel Economy and Lifetime Savings from Existing Conventional Technologies

	CAFE Rated Fuel Economy[a] (mpg)	Real World Fuel Economy[b] (mpg)	Fuel Economy Improvement vs. Baseline	Cost of Fuel Economy Improvement[a]	Lifetime Fuel Cost Savings[c]	Net Savings	Greenhouse Gas Savings (tons)	Avoided Toxic Emissions (lb.)	Smog Precursor Savings (lb.)
Small Car	48.4	38.7	57%	$1,125	$2,595	$1,470	30	16	35
Family Car	45.8	36.6	75%	$1,292	$3,590	$2,298	42	23	49
Pickup	33.8	27.0	61%	$2,291	$3,964	$1,673	46	25	54
Minivan	41.3	33.0	85%	$2,134	$4,534	$2,400	53	28	61
SUV	40.1	32.1	98%	$2,087	$5,346	$3,259	62	34	72
Fleet Average	41.8	33.4	74%	$1,693	$3,900	$2,207	45	24	53

a. Source: DeCicco, An, and Ross. *Technical Options for Improving the Fuel Economy of U.S. Cars and Light Trucks by 2010–2015.* Washington, DC. American Council for an Energy Efficient Economy, 2001.
b. CAFE fuel economy reduced by 20 percent.
c. Assumes a 15-year, 170,000-mile vehicle lifetime and a 5% discount rate. Average life based on scrappage rates from Davis 2000. Vehicle mileage based on 1995 National Personal Transportation Survey (NPTS) data.

bon dioxide reduction of -$49/ton of carbon dioxide avoided, in other words, consumers are paid to reduce their impacts on the environment while at the same time we are reducing our oil dependence.

Reaching 40 mpg by 2012

The UCS has compared the UCS/ACEEE fuel economy results with those from the recent National Research Council report and we find that the costs and improvements in fuel economy are very similar. . . . We estimate that a fleet fuel economy of 33 to 47 mpg could be reached at a retail price increase of about $1,700 to $3,800 per vehicle. This compares favorably to UCS/ACEEE estimates of a fleet fuel economy of 36–49 mpg at retail price increase of about $1,200 to $3,900. In both cases, consumers would be saving thousands of dollars at the gas pump. In most cases, this would be more than enough to pay for the cost of the fuel economy improvements, resulting in a net savings to consumers. . . .

The combination of both the UCS and the NRC results indicate that it is clearly feasible to reach a fleet average fuel economy of 40 mpg. We feel that such a standard could be phased in over 10 years, while the NRC analysis shows that similar fuel economy levels could be achieved within 10–15 years if weight reduction is not prominently used to reach improved fuel economy. In less than 10 years, both the NAS and UCS results agree that a fleet average of close to 35 mpg is technically feasible and cost effective.

The benefits to reaching a 49-mpg fleet by 2012 are quite significant. By 2012, we would have accumulated savings of 125 billion gallons of gasoline, this is about one full year's worth of gasoline. . . . In that same year, we would be saving about 1.9 million barrels of oil per day. This is more than the 1.7 million barrels per day we imported form Saudi Arabia [in 2000] and over three times the amount of oil we imported from Iraq. Consumers would also see significant benefits, with the US economy seeing net savings of 12.6 billion dollars in 2012 alone. On top of these financial benefits, over 40,000 new jobs would be created in the auto industry and close to 70,000 would be created in the US economy as a whole. In the end, increasing the average fuel economy of cars and trucks would both aid us in reducing our dependence on oil and help stimulate the economy.

Before the 40-mpg standards are phased in, UCS analysis indicates that average light truck fuel economy could be raised well above today's 20.7 mpg standard to that of cars (28.1 mpg) for about $670 in mass production. This increase in fuel economy could be achieved within 5 years using technologies available in cars today. By 2010, this increase in fuel economy would save 35 to 40 billion gallons of gasoline. . . . The overall benefit to consumers would be $7 billion dollars per year in 2010 alone and would be accompanied by significant reductions in greenhouse gas, toxic, and smog forming pollutants.

9

Raising Car and Truck Mileage Standards Will Endanger Motorists

Sam Kazman

Sam Kazman is general counsel of the Competitive Enterprise Institute (CEI), a libertarian think tank. He also heads CEI's Death by Regulation project, which focuses public awareness on the hidden dangers of government overregulation.

The federal government's fuel-economy program for new cars and trucks sets an average mile-per-gallon standard with which automobile manufacturers must comply. These standards, known as CAFE (for "corporate average fuel economy"), are currently set at 27.5 miles per gallon (mpg) for cars and between 20.7 and 22.2 mpg (by model year 2007) for light trucks and sport utility vehicles (SUVs). In order to comply with the standards, cars must be made lighter and smaller and are much more dangerous to their occupants. It is estimated that CAFE has contributed to tens of thousands of traffic deaths since it was instituted in 1975. Contrary to the arguments of the program's proponents, who routinely ignore this safety record, CAFE standards should not be raised any higher.

If the National Academy of Sciences discovered that a certain chemical was killing several dozen people a year, congressmen would rush to ban the substance. But in late July 2001, the Academy found that a federal policy was causing thousands of deaths—and now congressmen are scrambling not to end the policy, but to expand it.

Smaller cars equal more fatalities

The deadly policy in question is the federal government's new-car fuel-economy program, enacted in 1975 to reduce our dependence on foreign oil. Popularly known as CAFE (corporate average fuel economy), it requires the Transportation Department to set mile-per-gallon (mpg) stan-

Sam Kazman, "A Crashing Failure: The Stupid Tragedy of CAFE," *National Review*, vol. 53, September 17, 2001. Copyright © 2001 by National Review, Inc., 215 Lexington Ave., New York, NY 10016. Reproduced by permission.

dards. The passenger-car standard is currently 27.5 mpg, while the light-truck standard, which covers SUVs, is 20.7 mpg. A Republican proposal to make the standards slightly more stringent passed the House on August 1, [2001,] and the Democratic Senate may push for a far greater expansion after Labor Day.[1]

The real question is why anyone wants to expand CAFE at all. Two days before the House debate [in late July 2001], a panel of the National Academy of Sciences issued a long-awaited report, which concluded that CAFE had contributed to between 1,300 and 2,600 traffic deaths in a single year, as well as ten times that many serious injuries. Since CAFE has been in full force for over a decade, its cumulative human toll is probably in the tens of thousands.

By making cars smaller, CAFE has made them more dangerous.

By making cars smaller, CAFE has made them more dangerous. Larger, heavier cars are less fuel-efficient than similarly equipped smaller, lighter cars, but they're also safer. Research demonstrates that this holds true in practically every type of accident. Larger cars have more mass to absorb crash forces, and more interior space in which their occupants can "ride down" a collision before striking a dashboard or side pillar. For this reason, the smallest cars have "occupant death rates" more than twice as great as those of large cars.

Critics of CAFE have been raising this issue for years. A 1989 Brookings [Institution] Harvard [University] study estimated that CAFE caused a 14 to 27 percent increase in occupant fatalities—an annual toll of 2,200 to 3,900 deaths. A 1999 *USA Today* analysis concluded that, over its lifetime, CAFE had resulted in 46,000 fatalities. These findings are in the same ballpark as those of the Academy. CAFE's advocates, however, have uniformly claimed that CAFE isn't a factor in any deaths at all. A Sierra Club [environmental organization] brochure epitomizes this view, asking: "Can we improve fuel economy without sacrificing safety?" Its answer: "Absolutely. Long time safety advocates such as the Center for Auto Safety and [consumer-safety advocate] Ralph Nader support increasing CAFE to 45 mpg and point out that we can do so safely."

But Nader took a very different view back when large cars weren't as politically incorrect as they are now. In a 1989 interview on what type of car he'd buy, Nader said, "Well, larger cars are safer—there is more bulk to protect the occupant. But they are less fuel-efficient." Asked which cars are least safe, Nader replied: "The tiny ones." Clarence Ditlow's Center for Auto Safety took the same position. In 1972 it published a detailed critique of the Beetle, entitled *Small on Safety—The Designed-in Dangers of the*

---

1. In June 2002 the U.S. Senate voted to remove a provision from energy bill S. 517 that would have raised average car mileage requirements from 27.5 mpg to 36 mpg by 2015. In January 2003 the administration of President George W. Bush increased the government mandated standards for fuel efficiency in light trucks, a category that includes sport utility vehicles, from the current standard of 20.7 mpg to 22.2 mpg for model year 2007.

Volkswagen. Page after page explained how "small size and light weight impose inherent limitations" on safety. For example, the Beetle's compactness meant that "there is little space between the occupant and the windshield" and that "the gas tank is necessarily closer to the occupant than in larger cars."

Today, both Nader and Ditlow advocate higher CAFE standards.

Ignoring CAFE's record

The federal government, too, has ignored CAFE's safety record. CAFE is run by the Transportation Department's National Highway Traffic Safety Administration: You'd think an agency whose middle name is safety would be especially vigilant about CAFE's risks, but NHTSA has found those risks too embarrassing to confront. For over a decade, its basic position has been that CAFE is harmless.

The Competitive Enterprise Institute and Consumer Alert challenged that position in court. In 1992, a federal appeals panel ruled against the agency, harshly declaring that NHTSA had illegally "fudged" its analysis through a combination of "statistical legerdemain" and "bureaucratic mumbo jumbo." It concluded that "consumers who do not want to be priced out of the market for larger, safer cars, deserve better from their government."

[Although] clear evidence [has proven] that fuel economy is undermining safety, the politics of energy conservation trump the truth.

Three years later, after devising a new rationale for CAFE's innocence, NHTSA was upheld by another panel of judges. But that court still noted NHTSA's "failure to adequately respond" to the size-safety issue. Given the substantial judicial deference agencies generally receive, this comment indicated there was still something fishy about NHTSA's position.

Even when the consumer-safety establishment discovers clear evidence that fuel economy is undermining safety, the politics of energy conservation trump the truth. Last April [2001], Joan Claybrook's advocacy group, Public Citizen, released a new report on the Ford-Firestone tire fiasco, tracing the problem to Ford's attempt to boost the fuel economy of its Explorer SUV.[2] According to the report, Ford first requested that the tire's recommended inflation pressure be lowered to reduce rollover risk. That, however, raised its rolling resistance and worsened the Explorer's miles-per-gallon capability. To compensate, Ford asked for a lighter tire. Disaster ensued.

Public Citizen's report did not, however, prompt any rethinking of CAFE: One week later, Senator Dianne Feinstein sponsored a bill to raise SUV fuel-economy standards across the board.

Ironically, the popularity of SUVs is itself being used as an argument

2. In 2000 the Ford Motor Company was forced to replace thousands of defective Firestone tires installed on its Explorer SUV.

for higher CAFE standards. Higher standards, we are told, would lead to downsized SUVs, and fewer fatal mismatches in which subcompacts are demolished by invulnerable road monsters. But these grisly collisions are hardly typical of traffic accidents: The Insurance Institute for Highway Safety reports that fewer than 5 percent of small-car-occupant deaths occur in collisions with SUVs. In the Institute's view, "the high risks of occupants in light (and small) cars have more to do with the vulnerability of their own vehicles than with the aggressivity of other vehicles." If we're going to adjust the vehicle mix to improve safety, according to Institute senior vice president Adrian Lund, what we should be doing is "getting rid of the lightest cars on the road." CAFE, of course, is one of the main reasons these light cars exist in the first place.

A permanent trade-off between safety and fuel efficiency

CAFE's proponents argue that new technologies can raise fuel economy without further reducing the size of cars. They may be right, but CAFE will continue to constrain car weight regardless: No matter what fuel-saving technologies we put into the car of the future, adding weight to that car will both lower its fuel efficiency and increase its safety. The trade-off between these two factors will always remain. Dr. Leonard Evans, a renowned safety researcher and president of the International Traffic Medicine Association, says the CAFE proponents' argument is like "a tobacco executive claiming that smoking isn't risky because exercise and good diet can make smokers healthier."

Pro-CAFE-ers cite polls that supposedly demonstrate overwhelming public support for higher standards, but those polls don't mention the safety issue and the public itself knows little about it. The Competitive Enterprise Institute's polling, on the other hand, indicates that once the public learns of CAFE's actual death toll, only 19 percent still favor it.

In 1990, on the eve of the Gulf War [precipitated by Iraq's invasion of Kuwait], a coalition of CAFE advocates—including Ben & Jerry's [ice cream company] and [actor] Paul Newman—ran full-page newspaper ads claiming that a 3-mpg increase in the standards would make war unnecessary. "The price of gasoline," they intoned, "should never be a reason to send our sons and daughters off to die in a foreign war." That is correct, but at least in the Gulf War we knew that lives were being put at risk. CAFE advocates, on the other hand, have never admitted that their policy poses any risks whatsoever. The sad truth is that when it comes to blood-for-oil campaigns, CAFE ranks with the worst of them.

10

Oil Conservation Vouchers Will Strengthen America's Energy Independence

Martin Feldstein

Martin Feldstein is a professor of economics at Harvard University in Cambridge, Massachusetts.

The most effective way that America can reduce its dependence on imported oil is to encourage gasoline conservation through market-based incentives. Drivers who economize on gasoline should be rewarded. An effective means to achieve this outcome is the introduction of tradeable electronic Oil Conservation Vouchers. A limited number of vouchers would be distributed by the federal government to each household and company. These vouchers, whose value would be determined by the forces of supply and demand, would then be required to purchase gasoline. When motorists save on gasoline and do not have to use all of their vouchers, they can sell them for cash. Such a system provides an incentive to conserve gasoline. Furthermore, such a system would be superior to raising gasoline taxes, since it precludes excessive taxation and the government waste that accompanies it.

The United States can and should reduce its dependence on imported oil with the goal of achieving complete oil independence by 2020. Otherwise, we will continue to be hostage to the policies of the current and future rulers of Saudi Arabia, Iraq, Iran and their neighbors.

Reducing gasoline consumption is a priority

The U.S. now imports more than half of the oil it consumes. Experts predict this will be more than two-thirds within 20 years if there is no change in policy. A reformed regulatory approach to drilling and refining and a tax code that reflects the national-security importance of increasing domestic oil would encourage exploration and expand our refining

capacity. Both of those changes would help to slow our increasing reliance on oil imports. But the limited size of U.S. oil reserves means that
moving toward oil independence requires a substantial reduction in the
amount of oil that we consume.

One-third of our oil consumption is used to heat our homes. With
the right incentives, home heating could be converted over time to domestically produced natural gas and to electricity produced by a combination of nuclear power, coal, natural gas and renewable sources. The
national-security gain of reducing and eventually eliminating our oil dependence would outweigh the extra costs.

---

*The key to oil independence is to cut the amount of
gasoline that we use on the road.*

But the key to oil independence is to cut the amount of gasoline that
we use on the road. Major reductions in gasoline consumption are clearly
feasible. Over the past three decades, the quantity of gasoline per mile driven in automobiles fell 40%. Experts believe that existing technologies
can be developed further in the next decade to cut that gasoline per mile
to a half or even a third of today's level. If that is achieved, we will not
need to import oil.

Such new technologies will not be free. Cars that economize on gasoline will have more expensive engines and be built of more expensive materials. In the decades that it will take to shift the stock of cars to new and
much more fuel-efficient models, individuals can save gasoline by driving
less, using public transportation more, buying tires that increase fuel efficiency, and so on. The key is to find market-based incentives that will in-
duce individuals to make the right decisions.

Such market-based incentives are superior to the regulatory approach
that the government has used since 1975, when the Corporate Average
Fuel Economy (CAFE) standards began to force auto makers to produce
cars, but not trucks and other vehicles, that get more miles to the gallon.
The effect was to cause households to shift from cars to SUVs and light
trucks in a way that has caused the overall gas efficiency of new vehicles
as a whole to be no better now than it was a decade ago. Moreover, the
regulatory approach does nothing to encourage individuals to drive less,
to use their cars more efficiently, or to shift sooner to new and more fuel-
efficient vehicles.

The voucher system

We need to reward those who economize on gasoline and to make others
face the full cost of the gasoline they consume, including the implicit cost
of reducing national security by increasing our dependence on oil imports. An obvious but unacceptable way to reduce gasoline demand
would be a very big increase in the gasoline tax along European lines.
Raising the gasoline tax by a dollar a gallon would probably push the
price high enough to eventually eliminate the need for oil imports.

Some economists argue that an extra dollar a gallon would be justified even without considering national-security effects because it would make drivers face something closer to the true cost that their driving imposes on others by increasing congestion, road accidents, local environmental problems, and global warming. And the extra tax revenue of more than $100 billion a year could in principle be returned to the taxpayers by cutting income or payroll taxes in a way that strengthens work incentives enough to more than offset the adverse incentive effects of the gasoline tax itself.

But while all this might be true in principle, many Americans—including this economist—would be skeptical that the extra gasoline taxes would actually be returned by cutting other taxes. In Europe, the revenue raised by heavy gasoline taxes is used to finance a bloated welfare state. In the U.S., the current more limited gasoline tax finances government spending on roads and on urban transit systems. There is no hint of giving the money back to the taxpayers.

Fortunately, there is a better way to use market incentives to encourage gasoline conservation (and oil conservation more generally) without raising any tax revenue by using tradeable electronic Oil Conservation Vouchers. Here's how it might work for gasoline (with a similar system for home heating oil and other types of oil consumption).

A better way to . . . encourage gasoline conservation . . . without raising [taxes is] tradeable electronic Oil Conservation Vouchers.

If the government wants to reduce gasoline consumption in 2003 from a projected 180 billion gallons to say 140 billion gallons it would distribute 140 billion Oil Conservation Vouchers to individuals and businesses. Those who buy gasoline would pay the cash price at the pump plus one such "voucher" for each gallon of gasoline. The vouchers would not be pieces of paper but would be credits available in a debit account. The individual would spend these vouchers by using an Oil Conservation Voucher debit card in the gasoline pump just as a credit or bank debit card is currently used.

Because the vouchers are needed to buy gasoline, they would have a market value that is determined by the forces of supply and demand. If each voucher is worth, say, 75 cents, a driver would recognize that buying 10 gallons of gas means using up $7.50 worth of vouchers that could otherwise be sold for cash. The gas pumps could be programmed so that someone who lacks enough vouchers could buy them at the going price while anyone with excess vouchers could use them to offset some of the cash cost of the gasoline.

The political process would decide how many Oil Conservation Vouchers each household and firm would receive, taking into account geographic and demographic information. To the extent that this distribution of vouchers mirrors the income or payroll tax, the voucher system acts like a reduction in the individual's marginal income-tax rate just as it would if the government collected gasoline taxes and returned the revenue by cutting personal tax rates.

But there are three ways in which the tradeable voucher system is superior to a European-style gasoline tax. First, no revenue is collected and there is therefore no temptation to use the extra funds to increase government spending. Second, individuals enjoy a positive cash reward from selling excess vouchers if they economize on gasoline. And third, the government can accurately control the amount of gasoline consumed by the number of vouchers it issues, in contrast to the uncertain response to a higher gasoline tax.

Although we can reduce our oil imports only gradually over a substantial number of years, complete oil independence can be a realistic goal. We cannot change the fact that the Saudis and other Persian Gulf states have most of the world's oil reserves. But with the right policies, we can make that fact irrelevant.

11

Drilling in the Arctic National Wildlife Refuge Will Decrease America's Reliance on Foreign Oil

Walter J. Hickel

Walter J. Hickel served as U.S. secretary of the interior under President Richard Nixon from 1969 to 1974. He was elected governor of Alaska in 1966 and again in 1990.

The coastal plain of the Arctic National Wildlife Refuge (ANWR) in Alaska may contain enough oil to replace all oil imports from Saudi Arabia and Iraq, both unstable Middle East countries with links to terrorists, for a generation. Oil drilling in ANWR will not disturb the migratory caribou herd that congregates in the region from early fall to early May; the caribou will either coexist with the drillers or will simply move to Canada during the calving season. The techniques used to find and develop oil in Alaska are also highly advanced and will have little impact on the environment. Drilling for oil in ANWR will make America both safer and stronger economically.

The Senate Democrats have stubbornly refused to allow any oil exploration along the rim of the Arctic National Wildlife Refuge (ANWR) in Alaska.[1] Despite this latest vote, however, the issue is not going to go away. Given our continuing precarious dependence on overseas oil suppliers ranging from [former Iraqi president] Saddam Hussein to the Saudis to Venezuela's [Fidel] Castro-clone Hugo Chavez, sensible Americans will continue to press Congress in the months and years ahead to unlock America's great Arctic energy storehouse.

1. In April 2002 and again in March 2003, the U.S. Senate voted not to allow oil drilling in the Arctic National Wildlife Refuge (ANWR). As of summer 2003, Alaska senator Lisa Murkowski was attempting to authorize drilling in ANWR through a Senate budget resolution.

Walter J. Hickel, "ANWR Oil: An Alternative to War over Oil," *The American Enterprise*, vol. 13, June 2002, p. 54. Copyright © 2002 by the American Enterprise Institute for Public Policy Research. Reproduced by permission of The American Enterprise, a magazine of Politics, Business, and Culture. On the web at www.TAEmag.com.

I'm an Alaskan who believes the coastal plain of ANWR should be opened for intelligent exploration of its energy potential. ANWR is owned by all Americans. The very small portion of the refuge with oil potential can be explored and drilled without damaging the environment. At a time when America is dependent for vital energy supplies on overseas oil-producing countries, some of which are allied with terrorist groups, it makes no sense for us to ignore a region within our own borders that could supply up to a third of a trillion dollars worth of domestic energy—enough to replace completely all imports from Saudi Arabia or Iraq for a generation. There are already 171 million acres of land in Alaska fenced off for conservation and wilderness preservation. That's an area larger than the state of Texas.

ANWR's coastal plain, the only part of the refuge where oil is suspected to exist, is a flat and featureless wasteland that experiences some of the harshest weather conditions in the world. Temperatures drop to nearly -70 [degrees] F. There are no forests or trees. At all.

For ten months a year, the plain is covered with snow and ice and is devoid of most living things. Then, for a few weeks, a carpet of lichen and tundra emerges from beneath the snow. During that brief period, the migratory Porcupine caribou herd (named for the Porcupine River), one of Alaska's 20 caribou herds, may graze and calve on the plain. The animals seek breezes from the Beaufort Sea to help them cope with the blizzard of mosquitoes that hatch with the spring.

In 2001, the Porcupine herd didn't calve on the coastal plain. It gave birth to its young many miles to the east, across the Canadian border. It calved in Canada the previous year as well. There is nothing magical about the area.

It's unlikely that exploration and drilling on the coastal plain will harm the caribou. Most biologists expect the animals will react to the presence of human activity the same way the Central Arctic herd adjusted to oil development at Prudhoe Bay (the region to the immediate west of ANWR's coastal plain). That herd has not only survived, but flourished. In 1977, as the Prudhoe region started delivering oil to America's southern 48 states, the Central Arctic caribou herd numbered 6,000; it has since grown to 27,128.

It's unlikely that exploration and drilling on the coastal plain [of Alaska] will harm the caribou.

It is important to note that in the Arctic, oil drilling is restricted to the wintertime. And from early fall to early May, the Porcupine herd is not on the coastal plain at all. It roams south to the Porcupine Mountains and east into Canada.

ANWR covers an enormous area—nearly as much as New Hampshire, Vermont, Massachusetts, and Connecticut combined. The most beautiful sections of ANWR—8 million acres—are federally mandated wilderness areas where the only tolerated human activity is hiking, backpacking, camping, and rafting. No motorized vehicles are permitted, and no development of any kind is allowed. This wilderness heart of ANWR includes the mountains of the Brooks Range. Journalists often use images of these mountains when describing the coastal plain region and its rich energy supplies, but the Brooks Range will not be touched by development.

The key to America's energy future

When it set up ANWR, Congress recognized that the 1.5 million acre coastal plain possesses unique potential for large oil and gas reserves. It was stipulated that these resources could be developed at any time if Congress so voted. As a result, scientists have studied this area for more than 20 years, and their work has produced estimates of recoverable oil ranging up to 16 billion barrels. Most of these scientists recommend that exploration be allowed.

To compare how much petroleum may lie beneath ANWR, consider that the entire rest of the U.S. contains 21 billion barrels of recoverable oil. The monetary value of ANWR's pumpable oil is projected by the U.S. Energy Information Agency to be between $125 billion and $350 billion. This doesn't even count the region's vast natural gas potential.

This new source of Alaskan oil could more than supplant all of our annual oil imports from Saudi Arabia or Iraq.

How much would an oil reservoir that size, just a few miles from the already-built-and-paid-for trans-Alaska pipeline, mean to America and our energy future? The government estimates the coastal plain could produce 600,000 to 1,900,000 barrels of oil per day. This new source of Alaskan oil could more than supplant all of our annual oil imports from Saudi Arabia or Iraq and ensure that the trans-Alaska oil pipeline would continue to deliver domestically produced energy to American consumers for decades to come.

I have visited many oil-producing regions throughout the world. The production techniques are often primitive and risky, both for the workers and the environment. The technology used in Alaska's Arctic to find and develop oil is the best in the world. When and if development takes place on the ANWR coastal plain, there will be little traceable disturbance. Seismic tests to locate the oil, and the actual drilling after that, will take place in the winter, using ice roads that will melt later. Small gravel drilling pads, only six acres in size, will be used to tap vast fields and will be removed when drilling is complete. Alaska's "North Slope" oil workers take pride in challenging visitors to find any trace of winter work activities after the snow melts.

If oil is discovered in ANWR, the size of the surface area disturbed will be dramatically less than when Prudhoe Bay was developed 30 years ago. Experts estimate that less than 2,000 acres will be touched—out of the 1.5 million acres on the coastal plain, and the 19 million acres in ANWR as a whole.

The opposition to opening ANWR "isn't really economic, humanitarian, or even environmental. It is spiritual," wrote a *New York Daily News* columnist. "If all the oil in the refuge could be neatly sucked up with a single straw, the naturalists would still oppose it because [to them] human activity in a pristine wilderness is, in itself, an act of desecration."

That is an extreme philosophical position. America's access to energy is a serious national security issue. Over-dependence on foreign oil exposes us to energy blackmail and compromises our ability to protect our citizens and assist our friends in times of crisis. Our goal as Americans must be to produce as much energy as we can for ourselves. This need not undermine efforts to conserve energy nor undercut the push to discover alternate energy sources. We must extend the energy sources that are practical today, even as we pursue possible alternatives for the future.

Rather than shutting down the Alaska pipeline and our other Arctic oil infrastructure we should be linking them to the vast untapped resources that await us on ANWR's coastal plain. That will not only make America safer and stronger economically; it will provide the rest of the world with an environmentally responsible model of how to produce energy the right way.

12

The Arctic National Wildlife Refuge Should Remain Off-Limits to Oil Drilling

Edward J. Markey

Edward J. Markey is a Democratic U.S. congressman from Massachusetts.

Opening the Arctic National Wildlife Refuge (ANWR), a pristine wilderness area in northern Alaska, to oil exploration and drilling will have adverse environmental impacts and will not solve the problem of U.S. dependence on foreign oil. The section of ANWR presumed to hold oil reserves is a critical habitat for the Porcupine caribou herd; the industrial blight that accompanies oil exploration, such as toxic spills and chemical waste, may destroy the herd's habitat. Using existing technology to increase automobile fuel economy will prove much more effective at reducing dependence on foreign oil than domestic drilling, which will only reduce foreign oil dependence from 56 percent in 2001 to 50 percent in 2011.

O ne of the most magnificent wildlife reserves [Arctic National Wildlife Refuge (ANWR)] in America has been targeted for oil and gas development. It is threatened as never before, and will lose its wild, untrammeled character forever if we do not organize to fight this threat. Today, Representative Nancy Johnson (CT-R) and I are introducing the Morris K. Udall Arctic Wilderness Act of 2001,[1] with more than 120 cosponsors, Republican and Democrat, all united in their goal to preserve this precious wilderness in its current pristine, roadless condition for future generations of Americans.[2]

---

1. As of fall 2003, this bill remains in congressional committee.   2. In April 2002 and again in March 2003, the U.S. Senate voted not to allow oil drilling in the Arctic National Wildlife Refuge (ANWR). As of fall 2003, Alaska senator Lisa Murkowski was attempting to authorize drilling in ANWR through a Senate budget resolution.

Edward J. Markey, statement introducing H.R. 770, Washington, DC, February 28, 2001.

Protecting a bipartisan legacy

We have a bipartisan legacy to protect, and we take it very seriously. It is a legacy of Republican President [Dwight] Eisenhower, who set aside the core of the Refuge in 1960. It is a legacy of Democratic President [Jimmy] Carter, who expanded it in 1980. It is the legacy of Republican Senator Bill Roth [Delaware] Democratic Representative Bruce Vento [Minnesota], and especially Morris Udall [Arizona-Democrat], who fought so hard to achieve what we propose today, and twice succeeded in shepherding this wilderness proposal through the House of Representatives.

Now is the time to finish the job they began. Now is the time to say "Yes" to setting aside the coastal plain as a fully protected unit of the Wilderness Preservation System.

Allowing this industrial blight [from oil drilling] to ooze into the [Arctic] Refuge would be an unmitigated disaster.

Every summer, the Arctic coastal plain becomes the focus of one of the last great migratory miracles of nature when 130,000 caribou, the Porcupine caribou herd, start their ancient annual trek, first east away from the plain into Canada, then south and west back into interior Alaska, and finally north in a final push over the mountains and down the river valleys back to the coastal plain, their traditional birthing grounds. This herd, migrating thousands of miles each year and yet funneling into a relatively limited area of tundra, contrasts sharply with the non-migratory Central Arctic herd living near the Prudhoe Bay oil fields.

The coastal plain of the Refuge is the biological heart of the Refuge ecosystem and critical to the survival of a one-of-a-kind migratory species. When you drill in the heart, every other part of the biological system suffers.

The oil industry has placed a bull's eye on the heart of the Refuge and says hold still. This won't hurt. It will only affect a small surface area of your vital organs.

Nevertheless, the oil industry has placed a bull's eye on the very same piece of land that Congress set aside as critical habitat for the caribou. The industry wants to spread the industrial footprint of Prudhoe Bay into a pristine area.

Let's take a look at the industrial footprints that have already been left on the North Slope. Look at Deadhorse and Prudhoe Bay. They are part of a vast industrial complex that generates, on average, one toxic spill a day of oil, or chemicals, or industrial waste of some kind that seeps into the tundra or sits in toxic drilling mud pits. It is one big energy sacrifice zone that already spews more nitrogen oxide pollution into the Arctic air each year than the city of Washington, D.C.

Allowing this industrial blight to ooze into the Refuge would be an unmitigated disaster. It would be as if we had opened up a bottle of black ink and thrown it on the face of the Mona Lisa.

An unnecessary invasion

But why invade this critical habitat for oil if we don't have to? The fact is, it would not only be bad environmental policy, it is totally unnecessary. Here's why.

Fuel economy: According to EPA [U.S. Environmental Protection Agency] scientists, if cars, mini-vans, and SUVs improved their average fuel economy just three miles per gallon, we would save more oil within 10 years than would ever be produced from the Refuge. Can we do that? We already did it once. In 1987, the fleetwide average fuel economy topped 26 miles per gallon [mpg], but in the last 13 years [as of February 2001], we have slipped back to 24 mpg on average, a level we first reached in 1981.

Simply using existing technology will allow us to dramatically increase fuel economy, not just by 3 mpg, but by 15 mpg or more—five times the amount the industry wants to drill out of the Refuge.

Natural gas: The fossil fuel of the future is gas, not gasoline, because it can be used for transportation, heating, and, most importantly, electricity, and it pollutes less than the alternatives. The new economy needs electricity, and it isn't looking to Alaskan oil to generate it. California gets only 1 percent of its electricity from oil; the Nation gets less than 3 percent, while 15 percent already comes from natural gas and it's growing.

Alaska has huge potential reserves of natural gas on the North Slope, particularly around Prudhoe Bay and to the west, in an area that has already been set aside for oil and gas drilling called the National Petroleum Reserve. Moreover, we have significant gas reserves in the lower 48 [states] and the Caribbean. The coastal plain of the Refuge has virtually none.

Why sacrifice something that can never be recreated, this one-of-a-kind wilderness, simply to avoid . . . sensible fuel economy?

Oil not in the Refuge: The National Petroleum Reserve in Alaska has been specifically set aside for the production of oil and gas. It is a vast area, 15 times the size of the coastal plain, and relatively under-explored by the industry. Anything found there is just as close to Prudhoe Bay as the Refuge, but can be developed without invading a critical habitat in a national refuge.

In fact, just last October [2000], BP [British Petroleum] announced the discovery of a field in this Reserve that appears to be as large as Kuparuk, the second largest field on the North Slope. While the potential for oil in the Refuge still appears larger than in the Reserve, the Reserve holds much greater promise for natural gas, so that every exploratory well has a greater chance of finding recoverable quantities of one fuel or the other.

Our dependence on foreign oil is real, but we cannot escape it by drilling for oil in the United States. Energy legislation introduced in Congress [in 2001] attempts to set ambitious new goals for independence yet it would only reduce our foreign oil dependence from 56 percent today to 50 percent 10 years from now, which simply underlines the futility of trying to drill our way to independence.

We consume 25 percent of the world's oil but control only 3 percent of the world's reserves. Seventy-six percent of those reserves are in OPEC [Organization of Petroleum Exporting Countries], so we will continue to look to foreign suppliers as long as we continue to ignore the fuel economy of our cars and as long as we continue to fuel them with gasoline.

Sensible fuel economy should preclude domestic drilling

The public senses that a drill-in-the-Refuge energy strategy is a loser. Why sacrifice something that can never be recreated, this one-of-a-kind wilderness, simply to avoid something relatively painless—sensible fuel economy?

A 2001 poll, done by Democratic pollster Mark Mellman and Republican pollster Christine Matthews, shows a margin of 52 to 35 percent opposed to drilling for oil in the Refuge.

The public is making clear to Congress that other options should be pursued—not just because the Refuge is so special, but because the other options will succeed where continuing to put a polluting fuel in gas-guzzling automobiles is a recipe for failure.

Sending in the oilrigs to scatter the caribou and shatter the wilderness is what I call "UNIMOG energy policy." You may have heard about the UNIMOG. It is a proposed new SUV that will be 9 feet tall, 7½ feet long, 3½ inches wider than a Humvee [a type of SUV], weigh 6 tons, and get 10 miles per gallon.

That's the kind of thinking that leads not just to this Refuge, but to every other pristine wilderness area, in a desperate search for yet another drop of oil. And it perpetuates a head-in-the-haze attitude towards polluting our atmosphere with greenhouse gases and continuing our reliance on OPEC oil for the foreseeable future.

Now that our energy woes have forced us to think about the interaction of energy and environmental policy, it is a good time to say "NO" to a UNIMOG energy policy and "YES" to a policy that moves us away from gas-guzzling automobiles to clean-burning fuels, hybrid engines, and much higher efficiency in our energy consumption.

If we adopt the UNIMOG energy policy, we will have failed twice. We will remain just as dependent on oil for our energy future, and we will have hastened the demise of the ancient rhythms of a unique migratory caribou herd in America's last frontier.

We have many choices to make regarding our energy future, but we have very few choices when it comes to industrial pressures on incomparable natural wonders. Let us be clear with the American people that there are places that are so special for their environmental, wilderness, or recreational value that we simply will not drill there as long as alternatives exist.

The Arctic Refuge is Federal land that was set aside for all the people of the United States. It does not belong to the oil companies, it does not belong to one State. It is a public wilderness treasure; we are the trustees.

We do not dam Yosemite Valley [in Central California] for hydropower. We do not stripmine Yellowstone [National Park] for coal. We do not string wind turbines along the edge of the Grand Canyon.

And we should not drill for oil and gas in the Arctic Refuge. We should preserve it, instead, as the magnificent wilderness it has always been, and must always be.

Organizations to Contact

The editors have compiled the following list of organizations concerned with the issues debated in this book. The descriptions are derived from materials provided by the organizations. All have publications or information available for interested readers. The list was compiled on the date of publication of the present volume; names, addresses, phone and fax numbers, and e-mail and Internet addresses may change. Be aware that many organizations may take several weeks or longer to respond to inquiries, so allow as much time as possible.

American Council for an Energy-Efficient Economy (ACEEE)
1001 Connecticut Ave. NW, Suite 801, Washington, DC 20036
(202) 429-8873 • fax: (202) 429-2248
e-mail: info@aceee.org • website: www.aceee.org

ACEEE is a nonprofit organization that maintains that energy efficiency and conservation will benefit both the U.S. economy and the environment. The council publishes books and reports on ways to implement greater energy efficiency, including *Smart Energy Policies* and *Strategies for Reducing Oil Imports*.

American Petroleum Institute (API)
1220 L St. NW, Washington, DC 20005-4070
(202) 682-8000
website: www.api.org

API is a trade and lobbying organization for the oil and natural gas industries. The institute contends that greater domestic oil drilling can be conducted with minimal impact on the environment. API publishes numerous position papers on energy-related issues available through its website, including papers calling for lower gasoline taxes and fewer restrictions on offshore oil drilling.

The Brookings Institution
1775 Massachusetts Ave. NW, Washington, DC 20036-2188
(202) 797-6104 • fax: (202) 797-6319
e-mail: brookinfo@brook.edu • website: www.brook.edu

The Brookings Institution is a private, nonprofit organization that conducts research on economics, education, foreign and domestic government policy, and the social sciences. In articles published by the institute in its quarterly journal *Brookings Review*, researchers and scholars have taken the position that free-market forces, not military intervention, should form the basis of U.S. energy policy. Brookings also publishes several energy-related books through its publishing division, the Brookings Institution Press.

Cato Institute
1000 Massachusetts Ave. NW, Washington, DC 20001-5403
(202) 842-0200 • fax: (202) 842-3490
e-mail: cato@cato.org • website: www.cato.org

The Cato Institute is a libertarian public policy research foundation dedicated to limiting the role of government and protecting individual liberties. It disapproves of spending taxpayer funds to intervene in the political affairs of oil-producing countries, arguing that the world oil market will respond evenly to the forces of supply and demand and preclude oil shortages. The institute publishes the quarterly magazine *Regulation* and the bimonthly *Cato Policy Report*, both of which have featured articles critical of U.S. foreign policy as it pertains to oil in the Middle East.

Heritage Foundation
214 Massachusetts Ave. NE, Washington, DC 20002-4999
(202) 546-4400 • fax: (202) 546-8328
e-mail: pubs@heritage.org • website: www.heritage.org

The Heritage Foundation is a conservative think tank that advocates free enterprise and limited government. Heritage researchers maintain that, in order to reduce dependence on imported oil, the U.S. government must remove the regulations that prevent oil companies from drilling for oil in certain areas of the United States. Its publications include the quarterly *Policy Review* and on-line resources such as *Policy Research & Analysis*, which has featured articles on ways to strengthen national security by reducing dependence on oil from the Middle East.

Reason Public Policy Institute (RPPI)
3415 S. Sepulveda Blvd., Suite 400, Los Angeles, CA 90034
(310) 391-2245 • fax: (310) 391-4395
e-mail: feedback@reason.org • website: www.rppi.org

RPPI is a research organization that supports less government interference in the lives of Americans. Their libertarian philosophy stands firmly opposed to raising the fuel economy standards of automobiles, arguing that doing so will necessitate smaller cars and thus result in more fatal traffic accidents. The institute publishes the monthly magazine *Reason*.

Resources for the Future (RFF)
1616 P St. NW, Washington, DC 20036-1400
(202) 328-5000 • fax: (202) 939-3460
e-mail: webmaster@rff.org • website: www.rff.org

RFF is a think tank that conducts research to find ways of reducing U.S. dependence on foreign oil. It argues that alternative energies, not further domestic drilling, are the solution to future energy needs. On its website, RFF publishes numerous reports, lectures, and transcripts of congressional testimony provided by RFF researchers that promote environmentally sound energy policy for Americans.

Sierra Club
85 Second St., 2nd Fl., San Francisco, CA 94105-3441
(415) 977-5500 • fax: (415) 977-5799
e-mail: information@sierraclub.org • website: www.sierraclub.org

The Sierra Club is a grassroots environmental organization with chapters in every state. Maintaining separate committees on air quality, global environment, and solid waste, among others, it promotes the protection and conservation of natural resources. The Sierra Club contends that the Arctic National Wildlife Refuge in Alaska should remain off-limits to oil drilling, since drilling there would disrupt the habitat of animal species and the pristine Alaskan wilderness. It publishes the bimonthly magazine *Sierra* and the *Planet* newsletter, which appears several times a year, in addition to books and fact sheets.

Union of Concerned Scientists (UCS)
2 Brattle Square, Cambridge, MA 02238-9105
(617) 547-5552 • fax: (617) 864-9405
e-mail: ucs@ucsusa.org • website: www. ucsusa.org

UCS is a nonprofit alliance of scientists who contend that energy alternatives to oil and other fossil fuels must be developed to reduce pollution and slow global warming. The union advocates raising the corporate average fuel economy (CAFE) standards with which automakers must comply to forty miles per gallon by the year 2012. UCS publishes numerous articles and reports on alternative energy sources and ways to reduce fuel consumption, available on its website.

The United States Geological Survey (USGS)
12201 Sunrise Valley Dr., Reston, VA 20192
(888) 275-8747 • fax: (703) 648-4888
e-mail: webmaster@usgs.gov • website: www.usgs.gov

The USGS is a bureau of the U.S. Department of the Interior that provides scientific information on energy and fossil-fuel resources to the public. Through its website, the USGS provides numerous reports and forecasts on the future sustainability of domestic oil production.

Worldwatch Institute
1776 Massachusetts Ave. NW, Washington, DC 20036-1904
(202) 452-1999 • fax: (202) 296-7365
e-mail: worldwatch@worldwatch.org • website: www.worldwatch.org

Worldwatch is a nonprofit public policy research organization dedicated to informing policy makers and the public about emerging global problems and the complex links between the world economy and its environmental support systems. The institute takes the position that the U.S. government should not engage in wars for the sake of securing a steady oil supply, and it maintains that alternative energies need accelerated research and government support. It publishes the bimonthly *World Watch* magazine, the Environmental Alert series, and numerous policy papers.

Bibliography

Books

Rick Abraham   *The Dirty Truth: George W. Bush's Oil and Chemical Dependency*. Houston: Mainstream, 2001.

Anthony Cave Brown   *Oil, God, and Gold: The Story of Aramco and the Saudi Kings*. Boston: Houghton Mifflin, 1999.

Joanna Burger   *Oil Spills*. Piscataway, NJ: Rutgers University Press, 1997.

Kenneth S. Deffeyes   *Hubbert's Peak: The Impending World Oil Shortage*. Princeton, NJ: Princeton University Press, 2001.

Emirates Center for Strategic Studies and Research   *Gulf Energy and the World: Challenges and Threats*. London: I.B. Tauris, 2003.

Kate Gillespie and Clement Moore Henry, eds.   *Oil in the New World Order*. Gainesville: University Press of Florida, 1995.

Robert Gramling   *Oil on the Edge: Offshore Development, Conflict, Gridlock*. Albany: State University of New York Press, 1995.

Donald Paul Hodel and Robert Deitz   *Crisis in the Oil Patch: How America's Energy Industry Is Being Destroyed and What Must Be Done to Save It*. Washington, DC: Regnery, 1997.

Herbert Inhaber   *Why Energy Conservation Fails*. Westport, CT: Quorum Books, 1997.

Michael T. Klare   *Resource Wars: The New Landscape of Global Conflict*. New York: Metropolitan Books, 2001.

Ystein and Oaestein Noreng   *Oil and Islam: Social and Economic Issues*. Hoboken, NJ: John Wiley & Sons, 1997.

Nawaf E. Obaid   *The Oil Kingdom at 100: Petroleum Policymaking in Saudi Arabia*. Washington, DC: Washington Institute for Near East Policy, 2000.

Roger M. Olien and Diana D. Olien   *Oil and Ideology: The Cultural Creation of the American Petroleum Industry*. Chapel Hill: University of North Carolina Press, 2000.

Barbara E. Ornitz and Michael A. Champ   *Oil Spills First Principles: Prevention and Best Response*. New York: Elsevier, 2002.

John Paffenbarger   *Oil in Power Generation*. Paris: Organization for Economic Cooperation & Development, 1997.

Jeffrey Share and Joseph Pratt   *The Oil Makers: Insider's Look at the Petroleum Industry*. College Station: Texas A&M University Press, 2000.

U.S. Congress,
House Committee on
International Relations
Oil Diplomacy: Facts and Myths Behind Foreign Oil Dependency. 107th Cong., 2nd sess., 2002.

Periodicals

Gawdat Bahgat — "Oil and Militant Islam: Strains on U.S.-Saudi Relations," *World Affairs*, Winter 2003.

Donald L. Barlett and James B. Steele — "The Oily Americans: Why the World Doesn't Trust the U.S. About Petroleum," *Time*, May 19, 2003.

Mary H. Cooper — "Oil Diplomacy: Does the Need for Oil Drive U.S. Foreign Policy?" *CQ Researcher*, January 24, 2003.

Thomas J. Davis — "Should Auto Fuel-Economy Standards Be Tightened to Reduce Dependence on Foreign Oil?" *CQ Researcher*, February 1, 2002.

Robert Dreyfuss — "The Thirty-Year Itch: For Three Decades, Washington's Hawks Have Pushed for the United States to Seize Control of the Persian Gulf. Their Time Is Now," *Mother Jones*, March/April 2003.

Neil Ford — "The Crucial Oil Dimension," *Middle East*, May 2003.

Amanda Griscom — "Beyond Oil," *Rolling Stone*, November 28, 2002.

Danny Hakim — "Pitting Fuel Economy Against Safety," *New York Times*, June 28, 2003.

Thom Hartmann — "Dinosaur War: Iraq, the Oil Industry, and International Politics," *Ecologist*, December 2002.

Lisa A. Hayes — "Oil and Power," *OnEarth*, Winter 2002.

Donald F. Hepburn — "Is It a War for Oil?" *Middle East Policy*, Spring 2003.

Kevin Jost — "CAFE vs. Fuel Taxation," *Automotive Engineering International*, July 2002.

Michael T. Klare — "It's the Oil, Stupid," *Nation*, May 12, 2003.

Lutz C. Kleveman — "The New Great Game: Oil and Security," *Ecologist*, April 2003.

Marianne Lavelle — "Living Without Oil," *U.S. News & World Report*, February 17, 2003.

Amory B. Lovins and L. Hunter Lovins — "Energy Forever," *American Prospect*, February 11, 2002.

John V. Mitchell — "Fuel and Force," *World Today*, January 2002.

Romesh Ratnesar — "Do We Still Need the Saudis?" *Time*, August 5, 2002.

Mark Rubin — "Oil Drilling in Alaska: Domestic Production and the Environment," *Congressional Digest*, June/July 2001.

William Schneider — "War Has Its Reasons," *National Journal*, March 22, 2003.

Peter H. Stone — "Running on All Cylinders," *National Journal*, April 6, 2002.

Shibley Telhami

"The Persian Gulf: Understanding the American Oil Strategy," *Brookings Review*, Spring 2002.

Caitlan Thomas
and Burke Franklin

"Drilling in the Arctic National Wildlife Refuge," *Wildlife Society Bulletin*, Summer 2002.

John Zogby

"A War for Hearts and Minds: Most People in the Middle East Think the Iraq War Is About Oil," *New Scientist*, April 5, 2003.

Index